ALL ABOUT
STALLS & SPINS

BY EVERETT GENTRY

To my good friends Ken and Evelyn Cox

Everett Jo Gentry

TAB BOOKS Inc.

BLUE RIDGE SUMMIT. PA. 17214

FIRST EDITION

FIRST PRINTING

Library of Congress Cataloging in Publication Data

Gentry, Everett.
    All about stalls and spins.

    Includes index.
    1. Spin (Aerodynamics)   2. Stalling (Aerodynamics)
I. Title.
TL574.S6G46   1983   629.132'36   82-19410
ISBN 0-8306-2349-3 (pbk.)

Courtesy of Jim Larsen

Contents

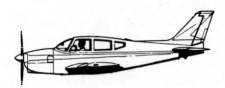

Acknowledgments

I am deeply grateful to my wife Sue for the many hours she spent typing the manuscript from drafts written in longhand. Too often this was like trying to read a stack of doctor's prescriptions. I owe a considerable debt to my daughter Suzann and to my son Ray for their faithful assistance.

I wish to thank the following persons for their critical readings of the manuscript and checking it for technical accuracy:

Major Damon L. Cooke, Chief, Flight Crew Training Branch of AWACS, and former Chief Instructor for E-3A pilot training at Tinker AFB, Oklahoma.

William J.G. McKim, Lt. Col. (AUS Ret.) of Midwest City, Oklahoma, a former technical writer for Philco-Ford Corporation.

Dan Stroud, an aerobatic instructor at Oklahoma City.

Bob Bishop, airshow pilot from Edmond, Oklahoma, who reviewed selected chapters.

Marion Cruce, a prominent soaring pilot from Oklahoma City, who reviewed Chapter 7 which pertains to sailplanes.

I wish to thank James S. Bowman, Jr. and his associates of the NASA Langley Research Center for their contributions to stall/spin research.

I am also indebted to John Paul Jones, former FAA Chief of Engineering and Manufacturing Flight Test Training in Oklahoma City, for letting me include a technical article written by him.

Thanks to Robert Gaines, Soaring Society of America, editor of "Safety Corner" of the SSA journal *Soaring,* for allowing me to use an article from that publication, and to Mary Raub of NTSB for her technical assistance. Mary formerly worked with me at FAA in developing the Private Pilot Airplane written tests.

I wish to express my profound gratitude to the following aircraft companies for permitting me to quote material from their publications and also for furnishing pictures of their aircraft:

Schweizer Aircraft Corporation
Cessna Aircraft Company
Beech Aircraft Corporation
Piper Aircraft Corporation
Mooney Aircraft Corporation

I wish to thank Duane Cole for allowing me to quote from several of his books, and also author John Lowery for the quotes I used from his book, *Anatomy Of A Spin*.

My thanks go to The Soaring Society of America for letting me quote from their training manual *Joy Of Soaring* by Carle Conway.

A special thanks to Jim Larsen of Kirkland, Washington, for the spectacular cover photo.

I must acknowledge the assistance I received from many of my pilot friends around the country. I feel their stall/spin stories are perhaps the most interesting and informative part of this book. The names of these individuals appear throughout the text.

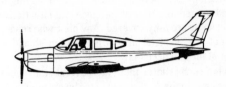

Introduction

Don't let the title of this book fool you. It was not intended to imply that I know all there is to know about stalls and spins. Frankly, I don't know anyone who does. The title does mean that all of this book is about stalls and spins and it contains everything I have been able to find out about them.

An article that I wrote entitled "To Spin or Not To Spin—That's The Question," appeared in the October, 1980, issue of *Private Pilot* magazine. TAB Books contacted me later and asked if I would be interested in expanding the article into a book on the same subject. Most of the material in the *Private Pilot* article is also in the book. The primary reason for writing on this subject is because there are some things that need to be said about stalls and spins that nobody has written to date.

In recent years many articles and books have been written that contained a multitude of stall/spin accident statistics. Of course, statistics can be used to glorify any point that someone is trying to prove. Therefore, there are no statistics, complex formulas, or equations mentioned in this book.

In addition to the stall/spin articles and books, several stall awareness and stall avoidance studies were conducted. These studies made recommendations for the reduction of accidents through innovation in ground and flight training curricula. But have they dealt with the *real* problem? In an effort to solve a problem, looking for answers in all directions except the right one is commonplace. One thing this book attempts to point out is the shortcomings of our current flight training. *What is being taught* as proper stall recovery procedures could be—and probably is—a causal factor in many stall/spin accidents.

What is not being taught concerns the lack of spin indoctrination training for private and commercial pilots, and in many cases flight instructors—a deficiency that can be corrected on several dual flights with an experienced instructor in an airplane certificated for spins. In talking to pilots, I find there is presently an unrealistic but deeply rooted fear of the unknown—a great deal of apprehension about spins. We are now reaping the results of 32 years of total abstinence from spin training. It is my hope that this book will stimulate the thinking of all general aviation pilots who read it, private and commercial pilots as well as flight instructors. Techniques and procedures to prevent close-call situations with unintentional stalls and spins are fully covered. Any pilot who reads, understands, and thinks about the material in this publication should cease being a candidate for a stall/spin accident.

At the end of each chapter there are several stall/spin stories. Some are my own, but most of them were told to me by my pilot friends around the country. I asked them to relate any experiences that may serve to deter others from similar mistakes or aircraft accidents. All of these pilots have been there. The personal pronoun "I" appears many times, but it does not always refer to me, since many of the first-person narratives were written by my friends.

Someone once told me that only three-quarters of what you read is true because even the honest writers usually oversell their subject. The opinions expressed in this book are my own. I have been thinking about them for many years and believe them to be true, yet my views may not be shared by all others in aviation. If there appears to be an angry note of impatience in this book, that is because it is long accumulated and difficult to restrain.

Chapter 1

What Happened to Stall/Spin Training?

The proposed cure for a problem does not always solve that problem. Under pressure from those who did not want to perform spins and from manufacturers of spin-proof airplanes, Civil Air Regulations Amendment 20-3 was adopted in 1949. This amendment eliminated spin training from pilot certification requirements and emphasized only stall training. The amendment stated in part: "(a) It will emphasize recognition of and recovery from stalls which, on the basis of available accident statistics, has proved to be the most dangerous maneuver to pilots; and (b) elimination of the required spin maneuver will act as an incentive for manufacturers to build, and operators of schools to use, spin-resistant or spin-proof aircraft."

The old Civil Aeronautics Administration Technical Manual 100 and the 1958 revised edition TM 106 entitled *Pilot Instruction Manual* reflected the changes in stall/spin training. Basically the same stall/spin subject matter with few changes was included in the 1965 edition of FAA Advisory Circular AC61-21, *Flight Training Handbook*. But the good things that were forecast—greater development and wide spread school use of spin-resistant or spin-proof aircraft—never came about.

An excerpt from the National Transportation Safety Board *Special Study—General Aviation Stall/Spin Accidents 1967-1969*, states the following:

"On the other hand, the evolution of spin-resistant or spin-proof aircraft has simply not been borne out. On the contrary, the trend toward modern-day, high-performance aircraft has resulted in spin characteristics considerably less favorable than those associated with predecessor aircraft. As the new generation of aircraft developed, compliance with the older, more stringent spin-recovery requirements became increasingly difficult and type certification spin tests for airplanes certificated in the normal category were, for all practical purposes, subsequently eliminated."

Most airplanes currently being manufactured are characteristically capable of spinning, and they are being flown by many pilots with no training or experience in spins or spin recovery procedures (Fig. 1-1). In recent years, many stall/spin articles have been written, along with several stall awareness and stall avoidance studies that were conducted. These studies made recommendations for the reduction of accidents through innovation in ground and flight training curricula. But, have they dealt with the real problem? Could it be that what is being taught and, equally important, what is *not* being taught concerning stall recoveries may be the *real* problem?

Fig. 1-1. Most aircraft are capable of spinning. Bottom photo shows J-3 Cub entering a right spin.

In the technical manuals and handbooks mentioned previously (and also in the Pilot's Operating Handbook for certain aircraft) there is, in my opinion, phraseology used that conveys the wrong message or creates in pilots a false sense of well-being concerning stalls.

AC61-21 states in part: "The use of ailerons in stall recoveries was at one time considered hazardous due to inefficient design of some older airplanes. In modern type certificated airplanes the normal use of ailerons will not have a detrimental effect in a stall recovery." If this statement is completely reliable, why is it possible in many modern-day airplanes to stall and enter a spin in the direction of a fully deflected down aileron, with the use of up elevator travel, and with both feet off the rudder pedals?

More misleading phraseology used in training manuals to define proper use of the controls in stall entry and recovery procedures is italicized below:

(1) Partial stall: " . . . with the recovery just after the break while the nose is falling, or *after the nose has fallen through the horizon*". (2) Takeoff-and-departure stalls: " . . . upon the student's first recognition of the stalls or after the stall has developed to such an extent that the *nose has pitched down through the level flight attitude*. In either case, recovery should be made by *relaxing the back pressure* on the elevator control and completed to straight-and-level flight by *coordinated use of all flight controls*." (3) "After the stall occurs, the recovery is made straight ahead *with the least loss of altitude*."

Let's take a look at several of these phrases to see how they might be interpreted by student pilots:

Words referring to recoveries	*Might convey this message*
" . . . or after the nose has fallen through the horizon."	Remain in a stalled condition, or delay the recovery.
". . . relaxing (or releasing) the back pressure."	Neutralizing the elevators is good enough.
" . . . with the least loss of attitude."	The important thing in stall recoveries is that you don't lose much altitude. Encourages timid use of recovery elevator pressure.
" . . . use of the ailerons will not have a detrimental effect in stall recovery." *or*	Always use the ailerons in stall recoveries because they are effective even if the

Words referring to recoveries	*Might convey this message*
". . . coordinated use of all flight controls."	wings are completely stalled. Forget the old wives' tale that said to do otherwise.

When an airplane is stalled and on the verge of rolling into an inadvertent spin, gentle recovery procedures will be about as effective as taking cough medicine to cure a critical case of pneumonia. Timid stall recovery procedures may be effective in docile side-by-side trainers, but how effective are they when the pilot of a fully loaded four to six-place airplane suddenly realizes the airplane is entering an inadvertent spin?

The desirable stall pattern of any wing is a stall which begins at the root section first. To overcome "tip stall" tendencies, airplane manufacturers use wingtip washout, fixed spoilers on the leading edge of the wing close to the fuselage, differential aileron travel (more up aileron deflection than down), slots, etc., yet each type wing platform has different stall characteristics. Also, by stalling one part of the wing before the other, the stall is not as abrupt as if the entire wing stalls at once.

Most of the present day side-by-side trainers use the rectangular wing planform, which is characterized by a strong root shall tendency that precedes the tip stall. This also provides adequate stall warning buffet and favorable stall characteristics.

It is true that relaxing the back pressure to recover from stalls in a docile trainer usually gets the job done. One of the prime reasons for this book is to show there would be fewer stall/spin accidents if we were teaching the use of "positive recovery elevator" or "brisk recovery elevator" to recover from every type stall in all airplanes. This does not mean that the forward pressure is required for an extended period of time. What it does mean is that positive elevator deflection can usually be released almost instantly, without the nose of the aircraft going any lower than normally occurs when timid recovery procedures are used.

Examine this excerpt from the 1965 edition of AC61-21, pertaining to recovery from a takeoff-and-departure stall in a climbing turn:

"Recovery from the stall can be effected, both with and without the use of power, upon the student's first recognition of the stalls or after the stall has developed to such an extent that the nose has pitched down through the level flight attitude. In either case, recovery should be made by *relaxing the back pressure* on the elevator control and *completed to straight-and-level flight by coordinated use of all flight controls.*"

That doesn't sound particularly urgent, does it? How effective will this procedure be in recovering from an unintentional stall of this type in an airplane with four to six persons aboard? A plain horse sense answer would be: "The airplane will get the bit between its teeth and be gone before the pilot realizes what has happened!"

This final quote from AC61-21 has probably been very effective over the years in conveying the message of gentleness or stressing timid use of recovery elevator when recovering from stalls. It states in part: " . . . In some planes a moderate action of the control column—perhaps slightly forward of neutral—is enough, while in others a forcible shove is required. *A reverse load thrown on the wings, however, may impede, rather than speed the stall recovery . . ."* The preceding italicized sentence has a lot of converts even though they may not understand its meaning. What does it mean? This statement would probably be true if a pilot was attempting to execute an English Bunt.

OLD CAA TRAINING MANUAL PROCEDURES

I have checked several older training manuals to determine how they covered stall and spin training. In reviewing the stall recovery procedures described in the *Civil Pilot Training Manual* (Civil Aeronautics Bulletin No. 23), I found very little emphasis placed on stall recoveries. This surprised me, as I instructed primary and secondary CPT during 1939-1941. The manual did stress that the ailerons were not to be used when practicing stalls. Here are two significant sentences from the publication concerning stalls:

"If it (the stick) is moved ahead too slowly, the recovery will not be clean and the airplane will simply 'mush.'

"If you ever sense imminent danger of a stall, THINK—with the stick forward."

I also reviewed the *Flight Instructors' Manual* (Civil Aeronautics Bulletin No. 5, revised 1941) procedures. This manual contained nearly identical information on stalls and spins to that which appeared in the Civil Pilot Training Manual. In fact, both manuals seemed to stress smooth use of the controls when entering practice stalls and relaxing the back pressure on recoveries. Ailerons were not to be used in stall recoveries. There was good coverage of spin training, which was a requirement at that time for private and commercial pilot certification.

The *Flight Instructors' Handbook*, CAA Technical Manual 105, devoted half of one page to a subject that continues to have a strong bearing on what is being practiced in stall recoveries. It states in part:

"In past years, it was customary to teach students to 'pop the stick' full forward for stall and spin recoveries. This was a wise procedure in early airplanes, especially those from the OX-5 days, whose stall recovery characteristics were sometimes uncertain. However, it is an unnecessary precaution in modern airplanes . . .."

When I was a kid I rode many times with pilots in OX-5 Wacos (Fig. 1-2) and Travel Airs (Fig. 1-3), seated in the front cockpit (with dual

Fig. 1-2. OX-5 powered Waco 10. (courtesy Omer Broaddus)

controls). We did lots of stall recoveries but I can't remember the stick ever being pushed full forward. The statement that it was customary to "pop the stick" *full forward* for stall recoveries was discussed with three members of the OX-5 Club. Each of these averaged about 2000 hours in various OX-5 powered airplanes, and they all agreed that this statement was heresay because they *never* pushed the stick full forward in stall recoveries. During spin recoveries, however, full down elevator was usually used in the same manner as that which is recommended for certain present-day airplanes. The paragraph in question that appeared in TM 105 (1956 edition) was deleted from later publications, yet it still influences the flying habits and thinking of pilots. Even today, to mention the expression "pop the stick" in stalls is considered sinful by many pilots. Those words have become dirty words. What is the big difference between saying *positive forward pressure, brisk forward pressure*, or *pop the stick*? Surely, some pilots would still be alive if they had popped the stick!

The development of new or revised editions of government manuals is often accomplished by cutting and pasting material from earlier manuals. A surprising number of the same paragraphs that appeared in the old CAA

Fig. 1-3. OX-5 powered Travel Air. (courtesy Omer Broaddus)

manuals forty years ago are still in the most recent manuals. The writer of a new manual is limited in making changes to established policies and procedures because these changes may be rejected by other persons who review the material before it is printed.

After a manual has been in use in the field for several years, many pilots and instructors memorize the procedures and parrot them as the gospel truth. Since government publications (including military manuals) are in the public domain and are not copyrighted, numerous companies reproduce and sell the manuals—often with a redesigned cover—as their own material. When aircraft manufacturers develop their owner's manuals and/or Pilot's Operating Handbooks, they also must conform with established policies and procedures for maneuvers.

This book deals primarily with stall training as presented in the 1965 edition of AC61-21, *Flight Training Handbook*. At the time of this writing (January 1982), the private, commercial, and flight instructor flight test guide procedures and maneuvers are referenced to the 1965 edition of this publication.

FLIGHT TRAINING HANDBOOK: 1980 EDITION

The Federal Aviation Administration published a revised 1980 edition, AC61-21A, *Flight Training Handbook*. This handbook, containing 17 chapters and 325 pages, is much more comprehensive and a great improvement over previous editions of the publication. It explains the recognition of stalls, fundamentals of stall recovery, use of ailerons in stall recoveries, stall characteristics, full stalls power-off, full stalls power-on, secondary stalls, imminent stalls power-on or power-off, maneuvering at minimum controllable airspeed, and accelerated maneuver stalls. The handbook also covers such demonstration stalls as the excessive back pressure (accelerated) stall, cross control stall, and elevator trim stall.

However, it does not list different methods of recovery for various types of stalls. The old 1965 edition contained phrases that told when to start the recovery from certain stalls. For example, "after the nose has fallen through the horizon" and similar phrases have been deleted from the new handbook.

The following excerpt from the 1980 edition of the Flight Training Handbook is a new look concerning the ailerons:

"Use of Ailerons in Stall Recoveries.

"Different types of airplanes have different stall characteristics. Most modern airplanes are designed so that the wings will stall progressively outward from the wing roots to the wingtips. This is the result of designing the wings in a manner that the wingtips have less *angle of incidence* than do the wing roots. Such a design feature causes the tips of the wing to have a smaller angle of attack than the wing roots during flight.

"Since a stall is caused by exceeding the critical angle of attack, the wing roots of an airplane will exceed the critical angle before the wingtips and, therefore, the roots will stall first. The wings are designed in this manner so that aileron control will be available at high angles of attack (slow airspeed) and give the airplane more stable stalling characteristics.

"When the airplane is approaching a completely stalled condition, the wingtips continue to provide some degree of lift and the ailerons still have some control effect. During recovery from a stall, the return of lift begins at the tips and progresses toward the roots. Thus, the ailerons can be used to level the wings.

"Using ailerons requires finesse to avoid an aggravated stall condition. For example, if the right wing dropped during the stall and *excessive* aileron control were applied to the left to raise the wing, the aileron deflected downward (right wing) would produce an even greater angle of attack (and drag), and possibly a more complete stall at the tip as the critical angle of attack is exceeded. The increase in drag created by the high angle of attack on that wing might cause the airplane to yaw in that direction. This adverse yaw could result in a spin unless directional control were maintained by rudder, and/or the aileron control sufficiently reduced.

"Even though excessive aileron pressure may have been applied, a spin will not occur if directional (yaw) control is maintained by timely application of coordinated rudder pressure. Therefore, it is important that the rudder be used properly during both the entry and the recovery from a stall. Thus, the primary use of the rudder in stall recoveries is to counteract any tendency of the airplane to yaw, or slip. The correct recovery technique then would be to decrease the pitch attitude by applying forward elevator pressure to break the stall, advancing the throttle to increase airspeed, and simultaneously maintaining direction with *coordinated* use of aileron and rudder."

HOW EFFECTIVE IS STALL TRAINING TODAY?

The following true story is an example of how inadequate some stall training might be.

A young man who holds a Private Pilot Certificate and has 150 hours flying time told me about the flight training he received prior to soloing in 1976. I have discussed flying numerous times with this pilot, who is in his late twenties. Although I have never flown with him, I found him to be very intelligent and I imagine he is above-average pilot material. He said that his first instructor was afraid of stalls and never permitted him to completely stall the airplane, but insisted that recovery controls be applied at the first sound of the stall warning horn. He also told me that the instructor never demonstrated a complete stall to him and he had never felt any stall buffeting, even though they were flying a Musketeer (Fig. 1-4) which was certificated for spins (Utility Category).

Sometime after solo, the pilot was assigned to another instructor. He told the new instructor that he would like to see "what a stall feels like," since he had never experienced one. The new instructor told him to try anything he wanted to because there was no attitude the airplane could be

Fig. 1-4. Beechcraft Musketeer.

put into without a safe recovery. They proceeded to execute stalls from every conceivable situation, and also recoveries from spin entries. This practice gave the pilot more confidence in the airplane and in his flying ability.

It is possible that some instructors never permit their students to completely stall an airplane, fearing the students may get them into a spin.

A CHANGE OF INSTRUCTORS

Many flight instructors continued to give their students spins after the requirement for spin training was deleted from pilot certification criteria in June of 1949. My good friend Col. Frank H. Dreher (USAF retired) of San Clemente, California, was an experienced flight instructor in civilian life for many years before entering the military. He tells of an experience that he had with a lady student, who was learning to fly at the time spin training was being dropped from the curriculum.

The lady student was nearly ready to solo, but when he started briefing her on spins she refused to go aloft for such training. Frank told her that he had never soloed a student without some introductory spin indoctrination and felt this was a safe and necessary procedure.

The lady then found a freelance part-time flight instructor who didn't consider spin training necessary or important. She soloed after two periods of takeoffs and landings. About three months later the part-time instructor spun in while circling at a low altitude over a friend's home, killing himself and a passenger. Perhaps more spin training with emphasis on unintentional entries would have been beneficial to this part-time instructor.

SPIN-RESISTANT OR SPIN-PROOF AIRPLANES

In the 1940s, several aircraft manufacturers were building airplanes that were supposed to be spin-resistant or spin-proof. One of these companies was Vultee, which produced the Stinson 105, a forerunner of the Stinson Voyager (Fig. 1-5). The 105 was a three-place high wing monoplane powered with a 75-hp Continental engine. This airplane had slots in the wings located in front of the ailerons, and another feature called *limiting pitch control power* (up elevator travel), which was supposed to make it difficult or impossible to completely stall the airplane. This airplane was built under earlier engineering requirements before the category system came into existence.

The flying service where I worked as a flight instructor acquired the Stinson dealership, and Vultee consequently sent a pilot to our airport to demonstrate the 105. The factory pilot was boasting that the airplane would not spin. After flying the demonstrator on a number of flights with prospective customers, I decided to see if it would spin.

Accompanied by a friend, I climbed the 105 to 4000 feet AGL and cleared the area below. A climbing turn to the left was entered, using a gradual buildup of left rudder and back pressure on the control yoke with some right aileron pressure also applied. Upon reaching full rudder and up elevator travel, the 105 spun beautifully for two turns and responded promptly when recovery controls were applied.

The demonstrator pilot said he did not believe that we had been spinning his airplane and wanted to be shown how we were able to make it spin. So he went with me to altitude and we spun it several times. Most

Fig. 1-5. Stinson 105 powered by a 75-hp Continental.

10

single-engine airplanes (built in the United States) that are difficult to spin will usually enter a spin from a climbing turn to the left due to the assistance of torque.

The demonstrator pilot returned to the factory and a week later we received a letter from Vultee. This letter stated that in rechecking the 105, they found the rudder and elevator stops were slightly beyond specification limits, which permitted excessive travel of those control surfaces. They assured us that after the adjustments were made on the control stops, it would be impossible for anyone to spin their demonstrator.

Several months later one of our customers took delivery of a new 105. Guess what—it was also a fine spinning airplane when entering from a climbing turn to the left.

It is probably true that the 105 or other airplanes that are difficult to spin may not enter a spin from a power-on or power-off stall straight ahead. Many pilots attempt to enter spins from a steep nose-high pitch attitude, but when the stall occurs, the nose drops in a large arc that permits the airplane to pick up some airspeed and enter a diving spiral rather than a spin. If these airplanes are stalled in a level flight attitude, or with the nose slightly above level flight attitude, they are more apt to enter a spin, when pro-spin rudder and elevator controls are applied.

Chapter 2

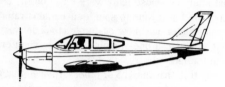

Certification of
Aircraft: Spin Testing

For many years private pilot flight training has emphasized teaching pilots to fly from point A to point B, and how to operate and comply with ATC procedures in the airspace system. I hope that some day we will get back to basics again and start teaching people to *fly*. Too many pilots today are 30-degree-pitch/60-degree-bank pilots; whenever they exceed either of those limits, they are operating beyond their realm of flight training.

In 1964 I was enrolled in a Pilot Flight Testing Procedures course at the FAA Aeronautical Center, and was amazed to see how shallow and gentle the lazy eights and chandelles were being performed. I had not performed either of these maneuvers in general aviation airplanes since 1951 because I had been working at a USAF contract school for about ten years. They told me the chandelles and lazy eights were being executed in this manner in order to comply with regulations and limitations for Normal Category airplanes. In other words, they would not be acrobatic maneuvers. It was also mentioned that some of the cleaner airplanes might go beyond the never-exceed-speed (V_{ne}) when attempting to perform steep lazy eights.

Let's take a look at the spin testing required for airplanes in the three categories (Normal, Utility, and Acrobatic). Federal Aviation Regulations spin testing certification requirements for general aviation airplanes are covered in FAR 23, which states in part 23.221: Spinning.

> (a) Normal category. A single engine, normal category airplane must be able to recover from a one-turn *spin* or a three-second spin, whichever takes longer, in not more than one additional turn, with the controls used in the manner normally used for recovery.

In addition—

(1) For both the flaps-retracted and flaps-extended conditions, the applicable airspeed limit and positive limit maneuvering load factor may not be exceeded;

(2) There may be no excessive back pressure during the spin recovery; and

(3) It must be impossible to obtain uncontrollable spins with any use of the controls. For the flaps-extended condition, the flaps may be retracted during recovery.

(b) Utility category. A utility category airplane must meet the requirements of paragraph (a) of this section or the requirements of paragraph (c) of this section.

(c) Acrobatic category. An acrobatic category airplane must meet the following requirements:

(1) The airplane must recover from any point in a spin, in not more than one and one-half additional turns after normal recovery application of the controls. Prior to normal recovery application of the controls, the spin test must proceed for six turns or three seconds, whichever takes longer, with flaps retracted, and one turn or three seconds, whichever takes longer, with flaps extended. However, beyond three seconds, the spin may be discontinued when spiral characteristics appear with flaps retracted.

(2) For both the flaps-retracted and flaps-extended conditions, the applicable airspeed limit and positive limit maneuvering load factor may not be exceeded. For the flaps-extended condition, the flaps may be retracted during recovery, if a placard is installed prohibiting intentional spins with flaps extended.

(3) It must be impossible to obtain uncontrollable spins with any use of the controls."

The maximum safe load factors (limit load factors) specified for airplanes in the various categories are:

Category	Permissible Maneuvers	Limit Load Factor*	
Normal	1. Any maneuver incident to normal flying.	3.8 Gs	−1.52 Gs
	2. Stalls (except whip stalls).		
	3. Lazy eights, chandelles, and steep turns in which the angle of bank does not exceed 60°.		

Category	Permissible Maneuvers	Limit Load Factor*	
Utility	1. All operations in the normal category. 2. Spins (if approved for that airplane). 3. Lazy eights, chandelles, and steep turns in which the angle of bank is more than 60°.	4.4	−1.76
Acrobatic	No restrictions except those shown to be necessary as a result of required flight tests.	6.0	−3.0

*To the limit loads given, a safety factor of 50% is added.

For Normal Category airplanes with a gross weight of more than 4000 pounds, the limit load factor is reduced.

Airplanes of older design (J-3 Cub, Aeronca 7-AC, Luscombe (Fig. 2-1), Cessna 120 (Fig. 2-2), etc., were built to requirements which did not provide for operational categories. Airplanes that do not have the category placard are designs which were constructed under earlier engineering criteria in which no operational restrictions were specifically given to the pilots. For airplanes of this type (up to weights of about 4000 pounds), the required strength is comparable to present-day Utility Category airplanes, and the same types of operation are permissible. For earlier design airplanes over 4000 pounds, the load factors decrease with weight so that these airplanes should be regarded as being comparable to the Normal Category airplanes designed under the category system, and they should be operated accordingly.

Fig. 2-1. Luscombe Silvaire. (courtesy Omer Broaddus)

Fig. 2-2. Cessna 120. (courtesy Omer Broaddus)

Small airplanes may be certified in more than one category if the requirements for each category are met. Some of the small four-place airplanes are certificated in the Normal Category with four persons aboard, and in the Utility Category when there are only one or two persons occupying the front seats.

REGULATIONS FOR ACROBATIC FLIGHT

FAR 91.71. No person may operate an aircraft in acrobatic flight—

(a) Over any congested area of a city, town or settlement;
(b) Over an open air assembly of persons;
(c) Within a control zone or Federal airway;
(d) Below an altitude of 1,500 feet above the surface; or
(e) When flight visibility is less than three miles.

FAR 91.15(c). Unless each occupant of the aircraft is wearing an approved parachute, no pilot of a civil aircraft, carrying any person (other than a crewmember) may execute any intentional maneuver that exceeds—

(1) A bank of 60 degrees relative to the horizon, or
(2) A nose-up or nose-down attitude of 30 degrees relative to the horizon.

DEFINITION OF ACROBATICS/AEROBATICS

The term *acrobatic* is used by FAA to define maneuvers involving unusual pitch or bank angles or acceleration. Part 91.71 of the Federal Aviation Regulations describes "acrobatic flight" as an intentional maneuver involving an abrupt change in the aircraft's attitude, an abnormal attitude, or abnormal acceleration not necessary for normal flight. Part 91.15 (c) refers to acrobatic flight as "an intentional maneuver that exceeds a bank of 60 degrees relative to the horizon or a nose-up or nose-down attitude of 30 degrees relative to the horizon."

15

Acrobatic is thus a more inclusive term than *aerobatic*, including not only spins, rolls, and loops but also steep turns, chandelles, or lazy eights (for example, whenever these exceed the indicated pitch or bank limits). The word *acrobatic* is usually used to refer to the stress applied to the aircraft, whereas *aerobatics* refers to the execution of a particular maneuver or figuration. Aerobatics is the word commonly used by pilots.

The maneuvers permitted by FAA for a given airplane depend upon the category of its certificate. A "fully acrobatic" aircraft must be designed to withstand the force of at least six Gs positive and three Gs negative. Such aircraft usually have special fuel systems which allow them to be flown inverted, either briefly or indefinitely. Fully acrobatic aircraft are usually without restrictions, except for certain limitations which may be placarded in the aircraft as a result of flight testing.

Aircraft in the "limited acrobatic" or "utility" category are stressed for 4.4 Gs positive and 1.75 Gs negative. Most of these aircraft are permitted all types of maneuvers except for sustained inverted flight.

In recent years five aircraft manufacturers have produced various models of aircraft certificated for acrobatics: Aerotek (Pitts), Beechcraft, Bellanca, Cessna, and Great Lakes. In addition, there are a number of aircraft no longer in production which are certificated as either fully acrobatic or acrobatic within certain restrictions. This includes many ex-military airplanes, antiques and classics, homebuilts, and other experimental airplanes.

An excerpt from FAA Advisory Circular 23-1, *Type Certification Spin Test Procedures*, states the following:

"A basic concept of type certification flight testing is to explore an envelope of the airplane's characteristics which is greater in all areas than the intended operational envelope. This is to assure that during normal operations, the operational pilot will not encounter any airplane characteristic that has not been explored by an experienced test pilot. With regard to the spinning requirements in CAR 3, type certification testing requires recovery capability from a one-turn spin while operating limitations prohibit intentional spins. This one-turn margin of safety is designed to provide adequate controllability when recovery from a stall is delayed.

"The spin requirements for Normal Category airplanes have changed over the years from six turns with a free control recovery to the present one-turn spin with a normal control movement recovery. Originally, and during the changes, there has never been any reference to the manner in which the spin entry should be conducted. The preamble of Amendment 3-7, dated May 3, 1962, states in part, 'These (one-turn spin) tests are considered to be an investigation of the airplane's characteristics in a delayed stall, rather than true spin tests.' This statement is significant and recognizes that CAR 3.124(a) does not require investigation of the controllability in a true spinning condition for a Normal Category airplane. Essentially, the test is a check of the controllability in a delayed recovery from a stall. Intentional and inadvertent, normal and accelerated stalls should be considered."

The emphasis placed on the recognition and awareness of stalls in training as a spin preventive has unquestionable merit. It would appear, however, that stall training alone, regardless of how rigorous, leaves something to be desired since such a complete dependence on avoidance of the stall leaves the results or the outcome of inadvertent spin entries highly questionable.

The first known airplane flight manual contained instructions issued with the 1911 Glenn Curtiss Pusher. Here is an excerpt from that manual concerning the landing approach and spins listed under the title "Rules Governing The Use Of Aeronautical Apparatus:" "Should the aeronaut decide to return to terra firma, he should close the control valve of the motor. This will cause the apparatus to assume what is known as the 'gliding position,' except in the case of those flying machines which are inherently unstable. These latter will assume the position known as 'involuntary spin' and will return to earth without further action on the part of the aeronaut." Some present-day pilots might consider this rule humorous, without realizing their own knowledge of spins is about on a par with the 1911 training course.

The following article regarding minimum requirements in aircraft design was written by John Paul Jones, former Chief of Engineering and Manufacturing Flight Test Training at the FAA Academy in Oklahoma City. This article appeared in the August 1975 issue of *Flight Operations* magazine, and it points out some of the problems associated with designing and building airplanes. George Haddaway, publisher of Dallas, Texas, has granted me permission to reprint the paper.

MINIMUM STANDARDS AND THE PILOT

"Among those concepts upon which the Federal Aviation Regulations are based, the one which appears to be least understood and generates the most questions is that of 'minimum requirements,' or 'minimum standards.' Since the idea is basic to all areas of aircraft design and operation, pilots should be clear as to its origin and its meaning.

"An aircraft structure differs in one respect from other structures. While it can be said that all aspects of any design are made final through compromise, this is the very essence of the design of an aircraft. If a question arises concerning the structural integrity of a proposed bridge or building, the matter can be settled by simply increasing the size of structural members or adding additional ones. The extra weight will hardly affect the efficiency with which the completed structure functions.

"An aircraft, on the other hand, has an additional and overriding criterion. Of course it must have structural integrity, but it must also have performance. And, in this area, weight is all-important. As a matter of fact, it is characteristic of an aircraft type which has a high pounds-per-horsepower ratio that, sooner or later, it will probably have structural problems from trying to take off through fences.

"Structural integrity, then, will be largely determined by the amount of structure carried. But this will also determine weight which, in turn, will determine performance. The final design must become a compromise between the strength and performance required for the operation in which the aircraft will engage.

"The compromises necessary in those areas are duplicated among numerous other aircraft capabilities. For instance, cruise speed can be increased by lowering profile drag through decreasing wing area. However, with this change comes increased stall speed and this will necessitate higher takeoff and landing values. These, of course, lead to longer field lengths and to controllability problems. Cruise speed may also be increased by substitution of more powerful engines—with higher initial cost, loss of range and increased operating expense. Similarly, stability and controllability must be balanced against each other, and each evaluated against the flight activity for which the design is intended.

"A continuous problem during the process of compromise is the possibility that the effort to strengthen one capability, or characteristic, may cause another to be weakened to the point that the needed safety level is not maintained.

"From the viewpoint of the FAR, then, all the foregoing considerations must be conducted within the overall parameter of maximum achievable safety. This creates a very real problem because compromise presupposes that something must be weakened in order for something else to be strengthened. The regulations, therefore, constitute an effort to ensure that no feature of aircraft design or operation exists.

"The concept of 'minimum standards' is a basic law within which the detailed regulations are developed. Today's FARs are authorized by the Federal Aviation Act of 1958, as amended, and this is the law of the land. That section of it which is concerned with aircraft design and operation is pervaded with the concept of safety. The Act recognizes that no one aircraft, or type, can have everything and that compromise will be necessary. It further takes note of the fact that these compromises can affect safety and it recognizes the need for an established level of safety below which builder and operator should not go (Fig. 2-3).

"The Federal Aviation Act, like many other such federal laws, does not establish detailed rules of procedure but empowers the appropriate agency (in this case the FAA) to develop and promulgate details. It does, however, specify in Title VI, Sec. 601, that: 'The Administrator is empowered and it shall be his duty to promote safety of flight of civil aircraft in air commerce by prescribing and revising from time to time: *Such minimum standards governing the design, materials, workmanship, construction, and performance of aircraft, aircraft engines, and propellers as may be required in the interest of safety* . . .'

Fig. 2-3. Below minimum standards: *"As we embark on another aerial adventure through fluffy clouds, over delicately hued mountains and misty vistas to far horizons, we are thinking of you"* reads humorous Christmas card sent in 1960 by Jim Dewey family. Dewey was an FAA GADO inspector at Van Nuys, CA. (courtesy Gordon Post)

"So, what is meant by the term *minimum standards?* In pilot's language it means that the FAA is charged with *establishing safety levels below which no further compromise is acceptable.* And, by its wording, this has to mean minimum standards for the design and operation of aircraft *to assure that they are safe.*

"The argument that a manufacturer can elect to go beyond the minimum standards set, and even has an obligation to do so, ignores basic laws of both physics and economics. By arbitrarily increasing a design's capabilities in one area the manufacturer must sacrifice somewhere else; in this process, he may either create a dangerous operational condition, or lower payload to the point of inadequacy or even bankruptcy.

"Throughout these comments we have ignored one most pertinent fact—a pilot can destroy any aircraft ever built, regardless of FAA design standards and manufacturer's care. For instance, a number of the early Model 35 Bonanzas lost wings in flight, and investigation showed that all were being operated under conditions beyond the pilot's capabilities. Almost all were weather accidents.

"In a classic case the late Mr. Shelby Kritzer (later of the Texas Aeronautics Commission) inadvertently found himself above an overcast and was forced to let down through it. His glide developed into a 'graveyard spiral' and he broke out about 500 feet above the treetops. In the following pullout the wings were 'sprung' approximately seven degrees. Since the plane was satisfactorily flown for about 100 miles and landed at its home field, a ferry permit was issued and it was flown on to the Beech factory for inspection. There the wings were proof-loaded and, in spite of their bent condition, were found still to have approximately 130 percent of their required strength. Even so, if the cloud base had been a little lower, the wings would have been pulled completely off. In a number of cases they were pulled off under what appeared to be identical conditions.

"We know exactly what occurred in this case because, when Kritzer was asked how the incident happened, he replied: 'It was 100 percent pilot error.'

"This pilot capability to emasculate all design safety efforts is not restricted to light, civil aircraft. A few years ago, during an Air Force demonstration, one of the supersonic fighter planes was decelerated at too slow a rate and therefore spent too much time in the transonic speed range. The massive tail structure was twisted and broken by the violent forces which came into play.

"Walter Beech once told the FAA in a certification meeting: 'Let's face it, we can't build a plane that somebody won't be able to tear up.'

"*Minimum standards* for design, then, result in the development of good aircraft, and 'minimum standards' for operation result in the world's safest flight operations. This is attested by the fact that United States-built aircraft are accepted and popular throughout the world, and are sought wherever pilots are free to choose. In addition, those countries which have developed aviation regulations of their own have borrowed heavily from the U.S. regulations and requirements. Finally, the ICAO (International Civil Aviation Organization) has established aviation laws which are used worldwide and these, also, are based upon U.S. regulations. Like the FARs and the regulations propounded by other free nations, the ICAO design and operation rules use the concept of minimum standards as a foundation for aircraft integrity."

AIRCRAFT SALESMAN IN TROUBLE

A fixed base operator recently told me that he knew of several accidents that occurred when aircraft salemen were demonstrating airplanes to customers.

Those stories go something like this: A private pilot/businessman, who has recently obtained an instrument rating and now considers himself an all-weather pilot, decides to buy a larger, better equipped airplane. He wants a four-place or six-place single-engine airplane or perhaps a good used light twin. After contacting a salesman at the local flying service that handles the make of airplane he is interested in, an appointment is made for a demonstration flight.

The aircraft salesman, also a private pilot without much flying experience, has been checked out in a number of different models sold by his company. The prospective customer arrives at the airport accompanied by his teenage son and a friend. In order to make a good impression, the salesman permits the prospective customer to fly the airplane from the left seat, while he occupies the right seat with dual controls. The prospect's son and his friend go along for the ride in the rear seats.

After climbing to altitude and trying a few turns, the prospective buyer requests permission to try some stalls and the salesman's reply is "Sure, go ahead." The pilot attempts several stalls using the same stall recovery technique that he learned in small docile trainers. Suddenly the airplane falls off on one wing and the nose points toward the ground. A spin is entered immediately, since the pilot's recovery action of pulling back on the yoke to raise the nose results in a fully developed spin. Realizing that something is wrong, the salesman closes the throttle. Further attempts to recover by the salesman, who has also never been in a spin before, are also futile. This story ends with another airplane-shaped hole in the ground.

After the spin was fully developed in this aft-CG loaded airplane, a successful recovery may have been impossible even with an experienced test pilot at the controls. Accidents of this type have possibly occurred more often than any of us realize. It is a case of two pilots being completely out of their realm of flight (headed straight down) in a Normal Category airplane that was not certified for spins.

UNINTENTIONAL SPIN IN THE TRAFFIC PATTERN

Here is a story of how a pilot's flying career might terminate in the final approach turn (Fig. 2-4). This pilot, named John, was an average weekend

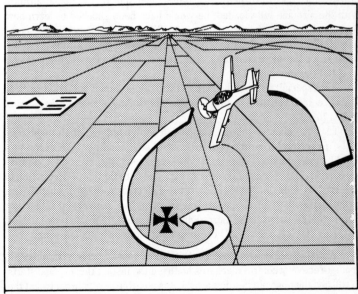

Fig. 2-4. How a crossed-control stall in the final approach turn results in a spin.

pilot with all previous experience limited to two-placed airplanes. John recently purchased a four-place airplane. The former owner gave him a quick demonstration flight/checkout. During the checkout, with only the two of them aboard the airplane, stalls were practiced, and the former owner tried to impress John with the airplane's stability and good stall characteristics. He was shown that the control yoke could be held all the way back in stalls, which resulted in buffeting and gentle porpoising. The owner proved the ailerons were effective by rocking the wings throughout the stalls.

Some time later, John and his wife invited another couple to join them for a weekend flight to a resort area. The flight was uneventful and upon arrival, John circled the small airport before entering the pattern. The runway was short. There was a 45° left crosswind. Hoping to land without overshooting the strip, John was flying too slow on the base leg with partial flaps extended. Failing to anticipate the rather strong cross-tailwind on base leg, he rolled into the base to final turn too late. In an effort to line up with the runway, he held left entry rudder, increasing it throughout the turn. To keep the bank from steepening he applied opposite (right) aileron pressure. As the airplane stalled the nose pitched downward and more back pressure was applied on the yoke. A crossed-control spin resulted.

There are several factors to consider concerning this typical crossed-control-spin accident:

☐ The weight of two extra people on board the airplane made it fly like a different animal, and the aft loading was the equivalent of having more up-elevator control. If the pilot had practiced stall recoveries at altitude with the airplane fully loaded, he would have been shocked to learn how different its stall characteristics were from those previously experienced.

☐ With partial flaps extended, the airspeed decays rather rapidly because of the increased drag. The pilot should have added power and adjusted the pitch attitude.

☐ Especially during the turn to final, pilots should get in the habit of asking themselves the question *"Am I holding rudder or crossing the ailerons in this turn?"*

☐ What about using the standard "coordinated use of all flight controls" to recover from the crossed-control stall condition? To many, this would mean applying right rudder pressure, additional right aileron pressure, and relaxing back pressure. The trouble with this corrective action is that the low wing (left) aileron is already deflected downward, which is one of the factors that caused the airplane to roll into an inadvertent spin to the left. In this situation, using more right aileron pressure would result in fully lowering the left aileron and creating more drag on a stalled wing.

Would it be possible to recover from this inadvertent spin entry? If prompt recovery control measures are used the instant the airplane starts to roll and the nose pitches downward, it is possible to regain control of the airplane. If faced with such a situation, I would briskly apply forward yoke

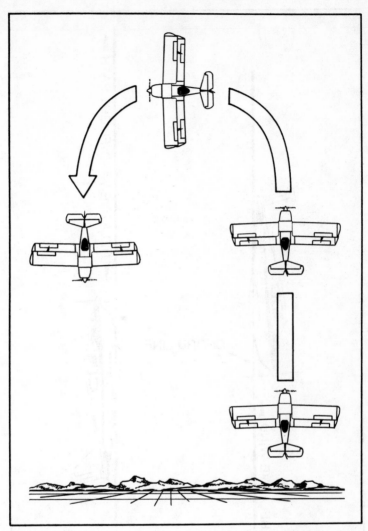

Fig. 2-5. Aerobatic pilots learn how to accomplish vertical flight into a hammerhead turn without entering a spin.

pressure, use enough right rudder to lower the nose straight ahead, neutralize the ailerons, and apply full throttle. This action would probably get the airplane flying above the treetops and coordinated use of all controls could be used to execute a go-around.

VERTICAL FLIGHT WITHOUT SPINNING

Figure 2-5 depicts an airshow pilot executing a hammerhead turn. Why is it possible for the pilot to fly the airplane in a vertical attitude without

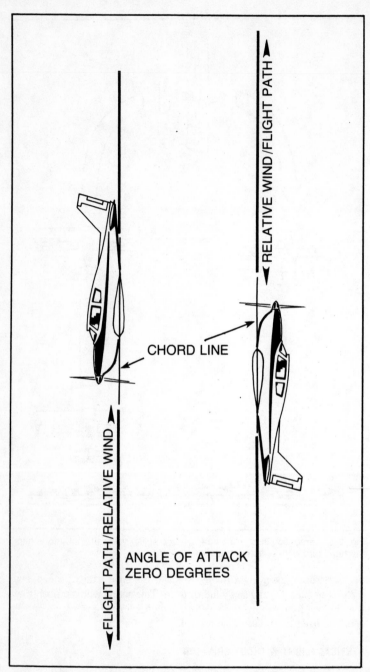

CHORD LINE

RELATIVE WIND/FLIGHT PATH

FLIGHT PATH/RELATIVE WIND

ANGLE OF ATTACK
ZERO DEGREES

Fig. 2-6. Vertical flight—zero angle-of-attack.

24

entering a spin, while some pilots experience unintentional spins in level flight? Perhaps the main reason is that the controls are never in a pro-spin position. Stated most simply, the maneuver is entered with considerable speed. As full power is applied, right rudder pressure is increased to correct for torque. The wingtips are clean (ailerons streamlined) in the pull-up. Before the airplane stalls, left rudder and forward elevator pressure are used to turn the airplane to the left (torque also assists in turning left). Since full left rudder is used over the top, some opposite aileron is necessary to keep the airplane from rolling over on its back. While the airplane is on its side almost motionless in a vertical bank, the wings are not stalled because there is no centrifugal force—no back stick pressure being applied. The airplane pivots about its vertical axis on the left wingtip. The nose (heavy end of the airplane) drops without entering a spin because pro-spin controls are not being used. The controls are coordinated because they are being applied in a manner that produces the desired results. In the vertical dive, the engine is throttled back as the airplane accelerates very rapidly. Smooth back pressure returns the airplane to straight and level flight headed in the opposite direction from which the maneuver was started.

I discussed "going over the top" in this maneuver with three professional aerobatic airshow pilots. Each of their airplanes is equipped with an inverted fuel system. A Citabria pilot and a Decathlon pilot told me they start applying rudder when the airspeed indicates around 40-45 knots. The Pitts pilot said that he is usually hanging on the prop with the engine wide open and the airspeed approaching zero as he starts feeding in rudder. Note the vertical flight angle of attack is zero (Fig. 2-6).

A word of caution to inexperienced acrobatic/aerobatic pilots: Stalling an airplane when it is headed straight up produces a very dangerous situation. The airplane will enter a tailslide and probably jerk the control stick from the pilot's hand. If the stick is all the way back during a tailslide, a violent whip stall will result. If the stick is all the way forward (elevators deflected down) during a tailslide the airplane falls upside down and backwards. Airplanes are not stressed to be flown in this manner, except for a few that are fully acrobatic.

While performing a hammerhead turn in a clipped-wing Cub at an airshow, the late Bevo Howard encountered a tailslide. He was able to land the airplane safely, however, though the trailing edges of both wings were bent downward around the rear spars as a result of falling upside down and backwards.

Tailslides are extremely dangerous and should be avoided at all costs, unless you are flying one of the few airplanes that are stressed for them.

Chapter 3

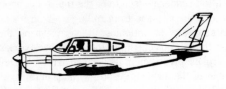

Back to Basics

To refresh your memory, some basic fundamentals and principles of flight are discussed in this chapter. They are not intended as a detailed discussion of maneuvers or a complete explanation of the elements of aerodynamics.

We hear a lot about the word *coordination*. What does it mean in aviation? Webster's Dictionary defines coordination as: "act of co-ordinating; state of being co-ordinate; harmonious adjustment or functioning."

The first definitions of this word that I learned are: "The ability to use and move the hands and feet together subconsciously and in proper relationship on the controls to produce the desired results"—or—"Using two or more of the controls in a manner that will produce the desired results." To fly your aircraft efficiently you must use the controls together. This is known as *coordination of controls* and is vital to smooth flying. Always remember: To get the desired results from the controls, you should think of using *pressures* rather than control *movements*. Rough and erratic use of all or any one of the controls will cause the aircraft to react accordingly, so it is important that you apply pressures smoothly and evenly. Proficiency is based on accurate "feel" and control analysis rather than mechanical movements. This valuable sense of feel can be developed in most pilots by directed practice in the correct performance of flight maneuvers. Obviously, the acquisition of the sense of feel and perfection of basic flight technique go hand-in-hand; one cannot be gained without the other. Here are several examples of coordination in flight.

Medium banked turns: Coordination of three controls (ailerons, rudder, and elevators) are required to execute a good turn as the aircraft must rotate about all three axes. After the bank has been established in a theoretically perfect medium-banked turn, all pressure on the aileron control may be relaxed. The ailerons will streamline themselves to the airflow when pressure on them is relaxed. The airplane will remain at the bank selected with no further tendency to yaw, since there is no longer displacement of the ailerons. At this point, pressure is also relaxed on the rudder pedals, and the rudder streamlines itself with the direction of the air passing it. If pressure is maintained on the rudder after the turn is established, the airplane will skid to the outside of the turn (Fig. 3-1). Holding opposite rudder pressure or entry aileron pressure will produce a slip to the inside of the turn.

Keep in mind that the rudder does *not* turn the airplane in flight. The airplane must be banked in a turn because the same force (lift) that sustains the airplane in flight is used to make the airplane turn. The airplane is banked and back elevator pressure is applied (coordination of the elevators). This changes the direction of lift and increases the angle of attack on the wings, which increases the lift. The increased lift acting perpendicular to the top of the banked wings (wingspan) pulls the airplane around the turn.

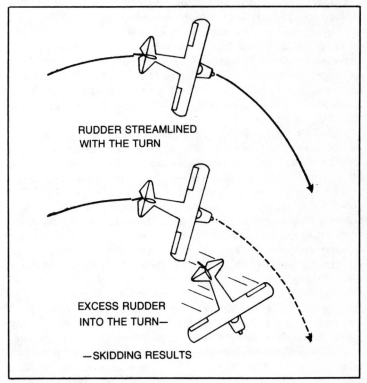

RUDDER STREAMLINED
WITH THE TURN

EXCESS RUDDER
INTO THE TURN—

—SKIDDING RESULTS

Fig. 3-1. Effect of rudder in turns.

What is the meaning of "crossed controls?" It is the simultaneous application of right rudder and left aileron or vice versa. In some maneuvers, it is necessary to cross the controls for them to be coordinated.

Coordination in slips: Slipping is done with the engine idling. Assuming that the airplane is in a glide (straight ahead), the wing on the side toward which the slip is to be made is lowered by the use of ailerons. Simultaneously, opposite rudder is applied to keep the airplane from turning in the direction of the lowered wing. The "forward slip" is a slip in which the airplane's direction of motion or flightpath continues the same as before the slip was entered. If the ailerons and rudder are used in proper proportion to one another and this produces the desired results, the controls are considered coordinated even though they are crossed. This is true of forward slips as well as of crosswind landings, when the touchdown is made on one wheel of the main gear.

Coordination in a slow roll: This maneuver requires aileron in the direction of the roll and opposite rudder at certain points of the roll. A considerable amount of forward stick pressure is required while inverted. Even though the controls are crossed throughout most of this maneuver in order to pivot on a point on the horizon, the controls will be coordinated if they are used properly to produce the desired results.

Coordination of throttle and elevator: While gliding in for a landing with the engine idling, if a pilot finds that he is undershooting and adds power, this action requires coordination of the throttle and elevator control. To accomplish this, advance the throttle smoothly to the desired power setting and simultaneously raise the airplane's nose to avoid gaining too much speed. As the throttle is retarded, the nose must be lowered again to the normal gliding attitude.

Consider these points about coordination: Suppose that using two pounds of aileron pressure and two pounds of rudder pressure produces a coordinated turn entry in a hypothetical airplane at cruising speed. Entering a turn in the same airplane at slow speeds may require a greater buildup of pressure on one control than on the other for them to be coordinated. The same may be true when entering a turn in this airplane at high speeds. It may also be necessary to use more pressure on one control than on the other during the recovery from turns at various airspeeds to prevent skidding or slipping. For example, at high speeds, the ailerons are more sensitive and need to be displaced less so there is less drag. You overcome aileron drag by using the rudder. At cruising speed in many airplanes, the rudder pressure necessary to overcome the aileron drag is about equal to the aileron pressure used. At lower than cruising speed, the rudder pressure must be greater than the aileron pressure. At higher than cruising speed, the required rudder pressure is less than the aileron pressure. Always use rudder and aileron pressure simultaneously, although the individual amount of pressure may differ depending on the effect of drag. Also, remember that aileron drag effect is present during recovery from a turn as well as during the entry.

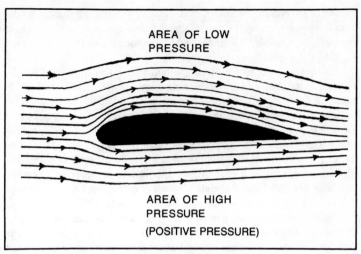

AREA OF LOW
PRESSURE

AREA OF HIGH
PRESSURE

(POSITIVE PRESSURE)

Fig. 3-2. Relative wind approaching an airfoil and pressures.

Here are some more definitions:

Lift: Scientist Bernoulli discovered that if the velocity of a fluid (air) is increased at a particular point, the pressure of the fluid (air) at that point is decreased. The airplane's wing is designed to increase the velocity of the air flowing over the top of the wing as it moves through the air, thereby decreasing pressure above the wing. This means the air on top must go faster. Hence, the pressure decreases, resulting in a lower pressure on top of the wing and a higher pressure below. The higher pressure then pushes (lifts) the wing up toward the lower pressure area (Fig. 3-2). It is this continuing difference in pressure that creates and sustains lift.

The airflow velocity over the top of the wing is greater than the velocity below the wing. This velocity increase can be visualized by realizing that the air must part and let the airfoil (wing) pass. Some of the air flows over the airfoil and some under the airfoil, but the airstreams must meet at the trailing edge of the wing. Since both airstreams flow around the airfoil in the same unit of time, the airstream with the greater distance to travel must have a greater velocity.

At the same time, the air flowing along the underside of the wing is deflected downward. Sir Isaac Newton's theorem states: "For every action there is an equal and opposite reaction." Therefore, the downward deflection of air reacts by pushing (lifting) the wing upward (Fig. 3-3).

The ability of a wing to generate lift is in proportion to the density of the air, speed, and the angle at which the wing strikes the air. Denser air produces more lift because the wing has a firmer substance to push against. More speed gives more lift because the wing can deflect more air downward in a given time.

Relative Wind: This is the motion of the air relative to the airfoil; it is parallel to and opposite the flight path. Relative wind during flight is *not* the natural wind, but is the direction of the airflow in relation to the wing as it moves through the air. The angle at which the wing meets the relative wind is called the *angle of attack.*

Angle of attack: This is the angle between the relative wind (or the flight path) and the chord line of the wing (Figs. 3-4, 3-5). The pilot controls the angle of attack with the elevators by moving the stick fore and aft. By changing the angle of attack, the pilot in effect changes the upper camber of the airfoil and the differential pressure. At high speeds where the wing affects a large amount of air, only a small angle of attack is needed to deflect the air a small amount. Conversely, at low speeds during which the wing affects a lesser amount of air, a larger angle of attack is needed to deflect the air a large amount. Thus, the angle of attack at various speeds must be such that the deflection of air is adequate for the amount of lift needed.

If the airplane's speed is too low, the angle of attack required will be so large that the air can no longer follow the upper curvature of the wing. It is forced to flow straight back, away from the top surface of the wing, from the area of highest camber. This causes a swirling or burbling of the air as it attempts to follow the upper surface of the wing because of the excessive change in direction. As the critical angle of attack is approached, the airstream begins separating from the rear of the upper wing surface. When the critical angle of attack is reached, the turbulent airflow which appeared first near the trailing edge of the wing (at lower angles of attack), quickly spreads forward over the entire upper wing surface (Fig. 3-6). This results in a sudden increase in pressure on the upper wing surface and a considerable loss of lift. Due to the loss of lift and increase in form drag, the remaining lift is insufficient to support the airplane, and the wing stalls. The critical (stalling) angle of attack is approximately 18 to 20 degrees on most airplanes.

Remember that the angle of attack is the angle between the chord line and the relative wind, *not* the chord line and the horizon. Therefore, an

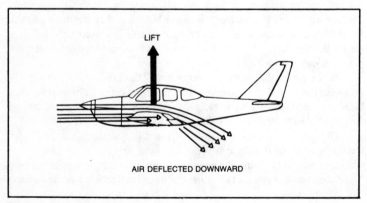

Fig. 3-3. Wing deflecting air downward.

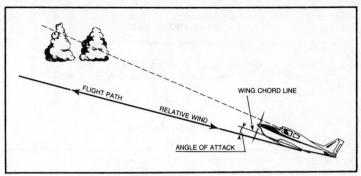

Fig. 3-4. It appears to the pilot above that he will pass beside the clouds, but his flight path is below the clouds.

airplane can be stalled in any attitude or at any airspeed by exceeding the critical angle of attack. A stall occurs when the pilot applies *abrupt or excessive back pressure* on the elevator control and exceeds the critical angle of attack.

It has often been stated that the most important factor in the art of piloting an airplane is the angle of attack and how it changes in flight.

Angle of incidence: This is the acute angle formed by the chord line of the wing and the longitudinal axis of the airplane. It is the angle at which the wing is attached to the fuselage, and should *not* be confused with angle of attack.

Camber: The curvature of the wing from the leading edge to the trailing edge. *Upper camber* refers to the curvature of the upper surface of the wing; *lower camber* refers to the bottom surface of the wing.

Center of pressure: If all the upward lift forces on the wing were concentrated in one place, there would be established a center of lift, which is usually called the *center of pressure* (CP). In addition, if all the weight of the airplane was concentrated in one place, there would be a center of weight, or as it is termed, *center of gravity* (CG). Rarely though are the CP and CG located at the same point. Most airplanes are designed to have their

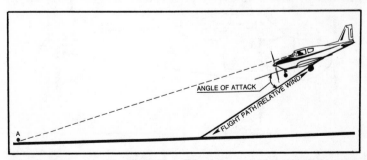

Fig. 3-5. It appears to this pilot that he will land near point A, but his flight path is much steeper.

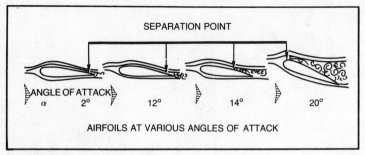

ANGLE OF ATTACK
α   2°   12°   14°   20°

AIRFOILS AT VARIOUS ANGLES OF ATTACK

Fig. 3-6. Effect of angle-of-attack on the separation point.

CG located slightly forward of the CP to create a nose-down tendency so the airplane will have a natural tendency to pitch downward and away from a stalling condition. This provides a safety feature in the characteristics of the airplane (Fig. 3-7).

Yaw: Rotation about the vertical axis is called yaw and is controlled by the rudder. This rotation is referred to as *directional control* or *directional stability.*

FORCES ACTING ON AN AIRPLANE IN FLIGHT

Lift: Lift is the upward force created by an airfoil when it is moved through the air. Lift acts upward and perpendicular to the relative wind and to the wingspan. The amount of lift generated by the wing depends upon several factors, including speed of the wing through the air, angle of attack, planform of the wing, wing area, and the density of the air (Fig. 3-8).

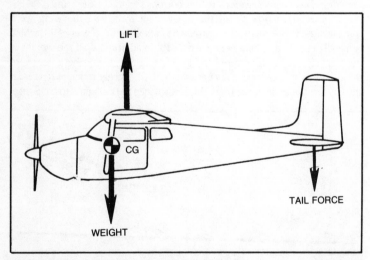

Fig. 3-7. The airplane is designed to nose down when the power is reduced, or when the airplane stalls.

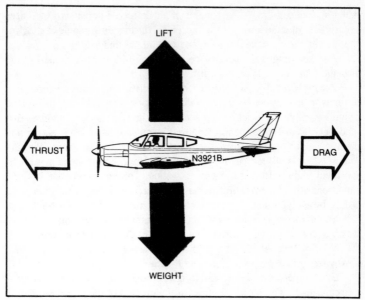

Fig. 3-8. Forces acting on the airplane in flight.

Weight (Gravity): Gravity is the downward force which tends to draw all bodies vertically toward the center of the earth. Lift does *not* always act opposite to weight. The airplane's center of gravity (CG) is the point on the airplane at which all weight is considered to be concentrated; it is a point of balance. For example, if an airplane was suspended from a rope attached to the center of gravity, the airplane would balance.

Thrust: The propeller, acting as an airfoil, produces the thrust or forward force that drives the airplane through the air. It receives power directly from the engine, and is designed to displace a large mass of air to the rear. It is this rearward displacement that develops the forward thrust that carries the airplane through the air. The thrust must be strong enough to counteract the forces of drag and to give the airplane the desired forward motion.

Drag: Drag is the rearward-acting force which resists the forward movement of the airplane through the air. Drag acts parallel to and in the same direction as the relative wind. Every part of the airplane which is exposed to the air while the airplane is in motion produces some resistance and contributes to the total drag. Total drag may be classified in two main types: induced drag and parasite drag.

Induced drag is the undesirable but unavoidable by-product of lift, and increases in direct proportion to increases in angle of attack. The greater the angle of attack up to the critical angle, the greater the amount of lift developed and the greater the induced drag. The airflow around the wing is deflected downward, producing a rearward component to the lift vector which is induced drag. The amount of air deflected downward increases greatly at higher angles of attack; therefore, the higher the angle of attack or the slower the airplane is flown, the greater the induced drag.

Parasite drag is the resistance of the air produced by any part of the airplane that does not produce lift.

Torque effect: Torque is a force, or combination of forces, that produces or tends to produce a twisting or rotating motion of an airplane. An airplane propeller spinning clockwise (as seen from the rear) produces forces that tend to twist or rotate the airplane in the opposite direction, thus turning the airplane to the left. Airplanes are designed in such a manner that the torque effect is not noticeable to the pilot when the airplane is in straight and level flight with a cruise power setting. Torque corrections in flight conditions other than cruising flight must be accomplished by the pilot. This is done by applying sufficient rudder to overcome the left-turning tendency.

Several forces are involved in the insistant tendency of an airplane of standard configuration to turn to the left. All of these forces are created by the rotating propeller. The four forces are *reactive force, spiraling slipstream, gyroscopic precession*, and *P factor*.

Centrifugal force: Centrifugal force is produced by any object moving in a circular path. The force acts toward the outside of the circle or turn. It reacts on an airplane during all turns, regardless of the plane of the turn. Centrifugal force, when acting on an airplane, usually acts in opposition to lift (Fig. 3-9).

The following explanations of other commonly used terms are excerpts from USAF *Aerodynamics For Pilots,* ATC Manual 51-3.

Accelerated stall: A stall with an airplane under acceleration, as in a pull-out. Such a stall usually produces more violent motions of the airplane than does a stall occurring in unaccelerated flight.

Adverse yaw: Yaw in the opposite sense to that of the roll of an aircraft; e.g., a yaw to the left with the aircraft rolling to the right.

Aerodynamic twist: The twist of an airfoil having different absolute angles of attack at different spanwise stations.

Center of pressure travel: The movement of the center of pressure of an airfoil along the chord with changing angle of attack; the amount of this movement, expressed in percentages of the chord length from the leading edge.

Differential ailerons: Ailerons geared so that when they are deflected, the up aileron moves through a greater angle than the down aileron, used to reduce adverse yaw or to lessen the control force necessary for deflection.

Downwash: A flow deflected or forced downward, as by the passage of a wing or by the action of a rotor or rotor blade.

Flat spin: A spin in which the airplane remains in a more level attitude than that of a normal spin, with centrifugal force holding the airplane away from the axis of the spin.

Laminar flow: A smooth flow in which no cross flow of fluid particles occurs, hence a flow conceived as made up of layers.

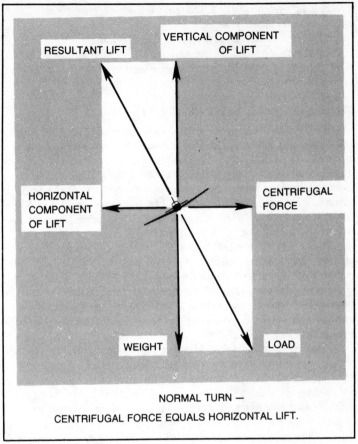

Fig. 3-9. Forces acting on an airplane in a turn.

Laminar flow airfoil: An airfoil specially designed to maintain an extensive laminar flow boundary layer from a body.

Negative G: The opposite of positive G. In a gravitational field, or during an acceleration, when the human body is so positioned that the force of inertia acts on it in a foot-to-head direction; i.e., the headward inertial force produced by a footward acceleration.

Positive G: In a gravitational field or during an acceleration, when the human body is so positioned that the force of inertia acts on it in a head-to-foot direction; i.e., the footward inertial force produced by a headward acceleration.

Spiral: A maneuver or performance, especially of an airplane, in which the aircraft ascends or descends in a helical (corkscrew) path, distinguished from a spin in that the angle of attack is within the normal range of flight angles; the flight path of an aircraft so ascending or descending.

Stalling angle of attack: 1. The minimum angle of attack of an airfoil or airfoil section or other dynamically lifting body at which a stall occurs; i.e., a critical angle of attack. 2. The angle of maximum lift.

Zero angle of attack: The position of an airfoil, fuselage, or other body when no angle of attack exists between two specified or understood reference lines.

DO YOU KNOW WHERE THE CONTROLS ARE?

Have you ever heard a TV news commentator say, "It's 10 o'clock; do you know where your children are?" Perhaps when pilots are flying at slow airspeeds, they might ask themselves this question: "Do you know where the controls are?" Too often pilots try to do the right thing with the wrong control. Misusing the ailerons to cover up for weak use of right rudder in correcting for torque is surely a causal factor in unintentional spin accidents. Many pilots are "aileron happy," using the rudder pedals only for foot rests.

Consider now the use of right rudder pressure to counteract the effect of torque during the performance of several maneuvers in a typical single-engine propeller-driven airplane.

Climb straight ahead: If the proper amount of right rudder pressure (or right rudder trim, if available) is maintained, the use of right aileron pressure should not be necessary. This is also true of the entry to stalls when practiced straight ahead.

Climbing turn to the left: Maintain right rudder pressure (or trim) throughout the turn. See if it is possible to keep the ball centered without using opposite aileron pressure; if opposite aileron pressure is required, use it sparingly.

Climbing turn to the right: Maintain right rudder pressure (or trim) throughout the turn. It probably will be necessary to use a slight amount of opposite aileron pressure to keep the ball centered, but don't use too much!

Steep turns: After the desired bank has been established and the airspeed drops below the cruising speed for which the airplane is rigged and trimmed, it will be necessary to apply and maintain right rudder pressure to compensate for torque. This is true in both left and right turns. The effect of torque is also increased when additional power is used. It is true that *slight* opposite aileron pressure usually is needed to control the overbanking tendency in steep turns, but it is wise to practice using as little opposite aileron pressure as possible. For example, while performing a steep turning to the left, some pilots use excessive opposite (right) aileron pressure to compensate for not using enough right rudder pressure to counteract torque. Aileron deflection has a profound effect on what the airplane will do when it is stalled in any of these maneuvers.

STEEP TURN SPIN-IN ACCIDENTS

One thing that is different about performing turns in an airplane and doing them in an automobile is that in an airplane, the entry aileron and rudder pressures are released (controls neutralized) after the bank has been established. In an automobile, the driver continues to hold pressure on the steering wheel in the direction of the turn until coming out of the turn, which is something a pilot needs to unlearn when flying an airplane.

While entering a turn, the airplane pilot has no problem with establishing the bank; however, blending the proper amount of back elevator pressure necessary in a steep turn is more difficult. The elevator control is the heaviest of all the controls, and the one that is not neutralized, but requires a pilot to maintain back pressure throughout the turn until starting recovery. In a 60 degree bank, the pilot must tighten the turn to 2 Gs in order to produce 1 G of vertical lift for the purpose of maintaining altitude (Fig. 3-10). The problem is that the pilot doesn't always know exactly how much back pressure to use, and too much or too little back pressure is likely to be applied, especially if the turn is entered rapidly or abruptly. When a steep turn is entered slowly and smoothly, a proper blending of back pressure can be used to maintain altitude as the bank steepens. Upon reaching the desired bank with the correct amount of back pressure, this pressure is maintained throughout the turn.

Have you ever rolled an airplane into a real steep banked turn—a bank that appeared to be about 80 degrees—and then pulled back hard on the yoke to maintain altitude? Many pilots have done this without realizing that the stress they were placing on the airplane's structure was close to or even beyond the yield point established for the airplane. At slightly more than 80 degrees of bank the load factor exceeds the limit of 6 Gs, the limit load factor of an acrobatic airplane (Fig. 3-11).

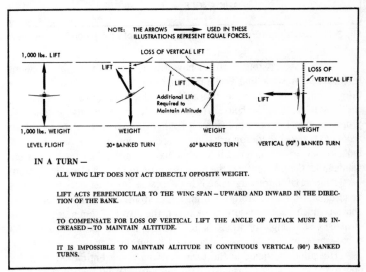

Fig. 3-10. Loss of vertical lift in turns.

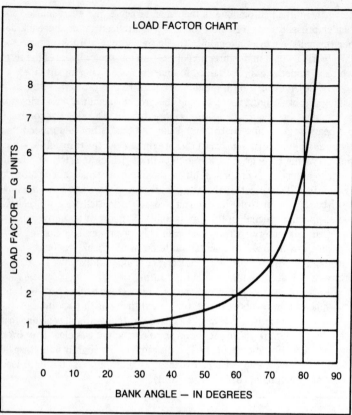

Fig. 3-11. Load factors produced at varying degrees of bank at constant altitude.

The following story portrays how a typical steep turn spin accident might occur: Assume that a pilot circles a friend's house at 500 feet above the ground, and in doing so, enters an abrupt steep left turn at a rapid rate of roll to stay close to the house. A considerable amount (rough estimate of proper amount) of back pressure is applied. To control the overbanking tendency, opposite aileron pressure is applied and held as the pilot then looks to see how he is doing in relation to the house. Glancing back at the windshield, the pilot notices the nose is a little too high and applies bottom (left) rudder. Of course, the nose is high because too much back pressure is being used. Now he has torque pulling left, bottom rudder, crossed ailerons, and too much back pressure working on the airplane. Let's assume this airplane's stall angle of attack is 18 degrees, and the pilot is in a 55-60 degree bank with an angle of attack of 16 degrees.

The pilot continues in the turn, dividing his attention between the house and nose of the airplane, until the airplane encounters a thermal (upward gust) which increases the angle of attack to 20 degrees. Suddenly the airplane stalls and rolls into a steeper left bank and the pilot thinks bumpy air caused the bank to steepen. More opposite (right) aileron is applied to raise the wing. The airplane's nose drops and the pilot tries to raise it with more back pressure, which instantly puts the airplane into a power spin to the left. Without a wind gust bringing about the stall, this pilot might have continued circling the house until entering propwash and turbulence created by his own airplane, but the outcome probably would have been similar since all the ingredients needed for spinning in were present. By referring to Fig. 3-12, you will find this airplane in a power-on clean configuration stalls with the wings level at 54 knots and at 76 knots in a 60 degree bank.

In an accident of this nature, there are several reasons why executing steep turns close to the ground is more hazardous than performing them at higher safe altitudes. Everything looks strange and different at the lower altitude, where the pilot tries to divide his attention between flying the airplane and circling an object on the ground. Additionally, he is often tense or under stress while attempting something that is considered dangerous or illegal. Performing low altitude steep turns is a perilous operation, especially when gradient winds are present or turbulence exists from wind blowing over uneven terrain or buildings.

Would it be possible for a pilot to recover from a situation like this without spinning in? At the moment the airplane stalls in the steep turn, if instant brisk forward elevator is applied to reduce the angle of attack, a recovery most likely can be accomplished as long as the bank has not

GROSS WEIGHT 2750 LBS		ANGLE OF BANK			
		LEVEL	30°	45°	60°
POWER		GEAR AND FLAPS UP			
ON	MPH	62	67	74	88
	KTS	54	58	64	76
OFF	MPH	75	81	89	106
	KTS	65	70	77	92
		GEAR AND FLAPS DOWN			
ON	MPH	54	58	64	76
	KTS	47	50	56	66
OFF	MPH	66	71	78	93
	KTS	57	62	68	81

Fig. 3-12. Stall speed chart.

exceeded 90 degrees. During such a recovery it would be wise to permit the airplane to dive some and gain speed *before attempting to level the wings*.

Recovery from this steep turn stall in which the critical angle of attack was exceeded requires prompt unloading of the wings by reducing the angle of attack. Since the airplane was not in a steep noise-high pitch attitude or at an exceptionally low airspeed, it should respond promptly to the recovery action without requiring a steep dive to regain airspeed. It was an accelerated stall that resulted from exceeding the critical angle of attack at a fairly high airspeed. In comparing the steep turn stall with a power-on stall straight ahead from a 45 degree nose-high pitch attitude, we find the latter stall occurs at a very low airspeed because of the steep pitch attitude, whereas in the steep turn, the airspeed is considerably higher. Recovery from a steep straight-ahead power-on stall requires lowering the nose more—usually below the horizon—to unload the wings and get them flying promptly.

Here is an important thing to remember about stalling in a turn at a low altitude: If recovery from the steep turn stall is delayed until the nose pitches downward in a spin entry, recovery before diving into the ground would be nearly impossible.

A LOW ALTITUDE STALL

My only experience with low altitude stalls occurred in 1936 at Bowman Field in Louisville, Kentucky, a few days after I received a Limited Commercial (L.C.) license. The L.C. license permitted a pilot to carry passengers for hire in the local area. My total dual and solo flight time was 63 hours.

The airplane in which this incident occurred was a two-place low-wing Aeronca powered by an 85-hp Leblond engine (Fig. 3-13). The wings of this airplane were full cantilever and tapered, and the airplane was a fixed-gear taildragger. The landing gear was mounted beneath each wing and covered with a streamlined fairing. On the underneath side of the fuselage between the landing gear struts, a flap or air brake panel was hinged to the fuselage. When not in use, this flap/air brake (approximately five feet in length and one foot wide) remained in the retracted position against the bottom of the fuselage.

The low-wing Aeronca was new, and I had received a 20 minute checkout before flying 15 minutes solo in the airplane. Several weeks later, an older private pilot friend, with three hours experience in the airplane, asked me to go flying with him in the Aeronca. He wanted to practice landings before taking it on a dedication tour of several small airports in Kentucky. The private pilot was in the left seat flying the airplane, and I occupied the right seat with access to the dual controls.

During the landing approach, my friend said "I'll show you how short I can land it using the flap/air brake." We were in a power-off normal glide when he lowered the flap, but he failed to lower the nose to maintain proper glide speed with the flap extended. Because of the great amount of drag this

Fig. 3-13. 1936 Aeronca low-wing, two-place, powered by 85-hp LeBlond five-cylinder engine.

surface produced, the airplane stalled within seconds and the airplane's nose pitched downward at a 40-45 degree angle. We were about 100 feet above the ground when the pilot realized we were falling. He said two words, and I can't repeat them here, but I believe they are the same two words that other pilots have uttered when they thought they were going to hit the ground. Frankly, it appeared to me that we going to register about seven on the Richter (seismograph) scale. As he removed his hands and feet from the controls, his next three words were: *"You got it!"*

I opened the throttle, added forward stick pressure and kept the nose pointed toward the runway in the 40-45 degree pitch attitude that the airplane had assumed. By the time we were ten feet above the runway, the airplane was flying again. I flared, closed the throttle, and landed.

Before flying this airplane solo again, my private pilot friend was given more dual instruction on stalls and landings by the owner.

In the case just mentioned, power saved the day. However, if the airplane had been entering a spin with the *nose straight down*, the use of power would merely have caused the airplane to strike the ground sooner than it would have with the throttle closed.

Chapter 4

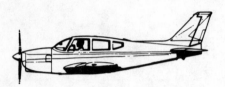

The Elevators Save People

"A pilot must have a memory developed to absolute perfection," Mark Twain once wrote. "But there are two higher qualities which he also must have. He must have good and quick judgment and decision, and a cool, calm courage that no peril can shake." Mark Twain was writing not of airplane pilots but of men who steered Mississippi River sternwheelers which often attained the terrifying speed of 14 miles per hour.

The nature of general aviation is such that most pilots are on their own once they receive a Private Pilot or Commercial Pilot Certificate, which means that they gain most of their later experience on a trial-and-error basis. Unknowingly, they acquire and practice bad flying habits and/or outright mistakes as time passes. It is not unusual for a general aviation pilot to find himself in situations where his experience level—and above all his flight training level—do not always provide for a proper response in an emergency. Pilots sometimes find themselves in a situation similar to the predicament of non-swimmers who fall into water over their heads, and with the same end results.

Consider the average pilot's stall training, which probably consists of practicing about 60 stall recoveries before receiving a Private Pilot Certificate. Stall training is usually practiced in small trainers that will recover from a stall themselves by merely releasing the controls. When the private pilot decides to step up to a four or six-place airplane, the checkout often consists of no more than in-the-pattern practice and perhaps six or eight stall recoveries. During the checkout with only two persons aboard the airplane, stall recoveries are not much different than those the pilot experienced in the docile trainer.

After checking out, this pilot may continue flying the larger airplane for a year or two on cross-country flights with no further stall practice. Everything goes well until one day when the airplane is accidentally stalled in a turn with a full load of passengers and baggage. It is quite possible that the pilot won't recognize the stall for what it is because the airplane has stalled when he isn't expecting it. When action is taken, however, the pilot's instincts—his most deeply established habit patterns of mind and body—cause him to revert back to the same recovery technique he learned practicing stalls in a docile trainer. If the nose pitches downward, he will, in a terrified effort to raise the nose and low wing, clamp the stick or yoke in the opposite rear corner, which converts a stall into a spin. Fortunately, this story as written is not true, but the plot is.

Too many pilots don't receive proper training and fly unaware of the possible extremes of stall behavior in a fully loaded airplane. Substantial differences in flying characteristics exist between various makes and models of airplanes, and also between various configurations and flight conditions of a given airplane. So often the pilot learns just enough to get up and down in an airplane. He also learns just enough to be dangerous.

The terms "brisk recovery elevator" or "brisk forward pressure" imply the use of some *down* elevator deflection, whereas relaxing the back pressure merely neutralizes or streamlines the elevators, and this may be too little and too late with some airplanes. In using brisk recovery elevator there is a definite movement of the elevators and usually this initial correction can be relaxed immediately, allowing the elevators to become streamlined. This does not mean that the airplane will dive excessively during the recovery from stalls nor is it necessary to lose any more altitude than would occur by merely relaxing the back pressure.

The important thing to accomplish in stall recoveries is to unstall the wings instantly and to regain flying speed. Many instructors are obsessed with placing too much emphasis on *minimum loss of altitude*. This technique encourages timid use of the elevators, which leads to inadvertent spins.

It seems wrong to have different methods for recovering from various stalls and flight regimes. Why not use positive recovery elevator on *all* stall recoveries, even approaches to stalls (partial stalls)? It is true that you can sneak out of some stalls on certain airplanes, but how effective will this method of using the controls be for recovering from the entry to an unintentional spin? When an inexperienced pilot accidentally stalls an airplane, this is not the time for him to be asking himself questions like these: "Let's see now, which recovery should I use? What kind of a stall is this? Do I relax the back pressure on this one, or should I hold the yoke back until after the nose falls through the horizon? Does this stall require positive forward pressure?" Stall recovery training should develop habits that will protect the pilot in the future when the airplanes being flown have entirely different and totally unexpected stall behavior.

I question the wisdom of having beginning students perform oscillation stalls (also called advanced stalls or rudder-controlled stalls) unless they have received spin or acrobatic training. Sooner or later, solo students attempting oscillation stalls will end up in a power spin. The one thing we don't want students to misconstrue is the use of full up elevator control as a means of recovering from stalls.

During the last 48 years, I have performed or ridden through many stalls and spins in many different airplanes. Some of these airplanes were prone to wingtip stalling, followed rather quickly by snapping into a spin, yet every one of them responded to prompt brisk use of recovery elevator. Deliberately stalling these airplanes in steep turns to the point where they were on the verge of a snap roll presented no recovery problem if positive, brisk recovery elevator was initiated. Simply getting pilots to be more elevator-conscious would undoubtedly reduce the number of stall/spin accidents.

One of the best spin prevention training exercises instructors can demonstrate is how to recover from the entry to an intentional spin by stopping the rotation instantly at any point. In other words, a normal spin entry is made with the throttle closed, using full up elevator travel and full rudder deflection. When the nose pitches down in the spin entry, apply brisk recovery elevator and enough rudder (opposite the direction of rotation) to make the nose drop straight ahead. The spin rotation stops immediately and the airplane can be returned smoothly to level flight with coordinated use of all controls. This has proven to be an excellent confidence-building maneuver for training military student pilots as well as civilian pilots.

The old Waco UPF-7 (open biplane taildragger) was a good example of how added tail-heaviness affects the stall characteristics of an airplane (Fig. 4-1). At high altitudes these airplanes could be stalled power-off to simulate three-point landings; as full up elevator travel was applied there was no tendency to roll or fall off to either side. During actual landing practice,

Fig. 4-1. 1941 Waco UPF-7 acrobatic trainer used in Secondary CPT program.

Fig. 4-2. Waco UPF-7 adjustable stabilizer.

however, when student pilots leveled off too high with this airplane, they often found that it would fall off to the left. If right aileron pressure was used as a corrective measure, the only thing accomplished would be a bent left lower wing aileron when it struck the ground.

Why did this airplane fall off to the left on landings and not have this tendency in power-off stalls? The difference was in the use of an adjustable stabilizer (lowering its leading edge) to trim the airplane for a power-off glide prior to landing (Fig. 4-2). When the Waco was taken to altitude and *trimmed* for a power-off glide, it did indeed fall off to the left when stalled. Using the adjustable stabilizer was the equivalent of having more up elevator travel, and therefore a more complete stall was attained. This is similar to what occurs with an aft CG loading of present-day airplanes when the rear seats are occupied.

It has been stated that modern aircraft generally have less elevator travel and do not stall as completely as older model airplanes. This is probably true *before filling the rear seats with passengers*.

It is true that well-designed ailerons are still effective at low airspeeds. Frise-type ailerons, wing slots, and differential aileron travel are some of the improvements in aileron design that have been in use for many years. It is my opinion that many of the present-day two-place side-by-side airplanes, which display good aileron control at high angles of attack, obtain this desirable characteristic more through limited up elevator travel than through an innovative aileron design. In other words, a tricycle-gear airplane doesn't need as much up elevator travel for landing as that required for a three-point landing in a taildragger. This means that aircraft with less elevator travel do not stall as completely. Limited up elevator travel closely resembles using only a small amount of back pressure on the yoke; consequently, the ailerons remain effective.

ENGINE FAILURE ON TAKEOFF

There are four persons aboard the airplane in Fig. 4-3, and the plane is climbing out after takeoff at best angle-of-climb speed. Upon reaching 400 feet AGL, the engine quits like it has been shot through the heart due to fuel starvation caused by operating on an empty fuel tank. During the pre-takeoff check the pilot was distracted by the passengers talking to him and he failed to check the position of the fuel selector. The unexpected occurrence of power failure on takeoff comes as a great surprise to pilots and many react as shown in the illustration—they just sit there without doing anything for a while. Gradually, the nose of the airplane drops some, or the pilot relaxes back pressure, and the airplane appears to be in a normal glide. But without establishing a glide at the proper speed the airplane will continue to mush with a high rate of sink until it strikes the ground.

If this pilot's training had included simulated forced landings on takeoff, and if the pilot had been taught to lower the nose briskly, flying speed would have been regained instantly. Further adjustments in the pitch attitude could then be made to maintain proper gliding speed. Any experienced flight instructor who has given many simulated forced landings on takeoff to students, recognizes the fact that most of them react the same as the pilot shown in the illustration.

At the moment the pilot of the airplane shown in Fig. 4-3 experienced power failure at 400 feet AGL, the airplane was not stalled. However, with a fully loaded airplane in a nose-high attitude, the airspeed decays rapidly and a stall will occur within a few seconds. With prompt, brisk lowering of the

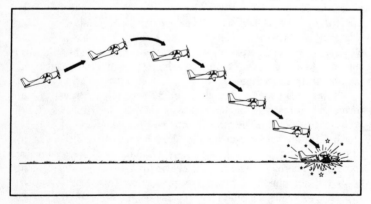

Fig. 4-3. An engine failure on takeoff at best angle-of-climb speed can result in a high rate of sink into the ground.

nose followed by prompt readjustment of the pitch attitude to regain normal glide speed, the airplane is flying rather than mushing at a high rate of sink. This permits the pilot to maneuver the airplane straight ahead to a landing with only a slight amount of turning to dodge objects in the flight path.

About actual forced landings someone once said, "If you can roll the wheels on the ground (or land gear up) with the proper airspeed, you might get skinned up a little but you won't get killed, because you will be *flying* and not *falling*."

When a four or six-place airplane is fully loaded, the pitch attitude at best angle-of-climb speed is usually moderately steep. On the other hand, if there are only two persons on board occupying the front seats, the pitch attitude at best angle-of-climb is quite steep and the visibility forward over the airplane's nose is limited. In either case, when engine failure occurs, prompt action is mandatory.

Someone may ask, "Suppose the power failure occurs at 50 feet above the ground. Won't the pilot dive into the ground?" The answer to this question is *no*. By lowering the nose immediately—before the airspeed decays—there is good control response and the airplane feels solid and normal, so the pilot can then flare for a landing. Failure to lower the nose at 50 feet AGL would probably result in either the airplane hitting the ground hard enough to wipe out the landing gear, or in a stall followed by the nose pitching downward at a steep angle.

Very few pilots get in trouble stalling and spinning in from straight gliding flight. Pilots most often get killed by losing control of their airplanes while turning, usually close to the ground with the engine operating normally. Most airplanes are very forgiving of the pilot when they are not in a turn! When airplanes spin in it is almost always out of the *bottom* of a skidding or crossed control turn. During a stall in a turn where the airplane goes over the top, pilots have a better chance of recovering as the wings pass through level flight.

A 1959 edition of the USAF Primary Flying student textbook for the T-28 trainer speaks briefly and clearly about what to do in case of engine failure on takeoff. It states: "CAUTION: If your engine should fail on takeoff, there is only one thing to do—land straight ahead. You may make a slight turn to avoid obstructions, but *never* try to turn back to the field." The T-28 shown in Fig. 4-4 was being flown by a pilot who complied with these instructions when the engine quit on takeoff.

For the three year period 1967 through 1969, an NTSB study revealed there were 238 general aviation stall/spin accidents that occurred during the takeoff and initial climb phase of flight. (I'm sorry, I promised to not mention any statistics.) Perhaps the single greatest stall/spin hazard comes from steep banks and excessive maneuvering in turbulent air at low altitudes. Without a stall, a spin cannot occur.

There has always been an unwritten law at airports throughout the country which most pilots strictly adhere to. It states: "*Never* attempt to turn back to the field if the engine quits on takeoff." I have known several

Fig. 4-4. When the engine on this USAF T-28A "shelled its goodies" on takeoff, a landing was made straight ahead gear-up with only minor damage.

pilots who were killed trying to turn back. Even if someone has gotten away with it, they would be looked down upon and other pilots would refuse to fly with them without dual controls. There were never any sanctions imposed on anyone for trying because the few who tried to turn back were usually killed in the attempt (Fig. 4-5).

One time when I was on takeoff in a T-6, oil quickly covered the windshield, and flames shot out of the exhaust manifold on the right side as far back as the front cockpit. The tower gave me permission to land immediately on the closest inactive runway. Of course, this was *not* a case of engine failure but it was necessary to get back on the ground as quickly as possible. After landing on the inactive runway, I shut the engine off. The mechanics soon discovered that the oil dilution system had malfunctioned and filled the crankcase with gasoline.

During my instructing days I always insisted that my students promptly lower the nose enough to make both of us light in the seat when they were given a simulated forced landing on takeoff during the climbout. Is there really anything wrong with momentarily going to zero Gs in a situation like this if a normal glide is established immediately thereafter?

A standard operating procedure used for years in the old CPT training, as well as in military primary flying schools, was to make the first turn after takeoff a level turn. This gave a built-in safety factor in case of engine failure at that point, and it also afforded the pilot a better opportunity to clear the area before turning. Most airplanes making the first turn a climbing turn will have more blind spots.

DON'T LET THE AIRPLANE FLY YOU

Instructors often tell their students, "*You* fly the airplane, don't let *it* fly you." There are several ways that inexperienced pilots get involved in serious accidents because they let the airplane fly them instead of the other way around. The following are examples of what I believe frequently happens and are causal factors in stall/spin accidents.

On takeoff: I remember a case where a student during preflight failed to check and adjust the stabilizer on a Piper PA-11 Cub before a solo flight. The Cub had been flown solo by another pilot prior to this student's flight, so the stabilizer was left in the power-off glide (landing) position, making the airplane very tail-heavy at full power. This student was flying solo from the front seat and when he got the throttle fully open the airplane jumped off the ground and assumed a 45 degree climb angle. It continued in a steep climbing turn to the left until reaching 75-100 feet AGL altitude, then stalled, fell off to the left, and hit the ground. The PA-11 was back on the

Fig. 4-5. Baby Ace pilot ran out of air while attempting to turn back to field after engine quit on takeoff—no injuries to pilot except skinned nose and bruised knee. (courtesy Ken Moore)

ground before the student realized what had happened, as he was just along for the ride and letting the airplane fly him. Fortunately, he was not hurt, but the only salvagable parts of the PA-11 were the rudder and tailwheel, and the aircraft logbooks which were in an office file cabinet.

To make certain that this would never happen to my students, I always had them fly the traffic pattern twice (dual) with the airplane out of trim. First, we would fly the airplane around the pattern leaving the elevator trim (or stabilizer) in a very tail-heavy position. The entire pattern was flown this way, but I always kept my hand close behind the stick because even the sharp students were apt to let the airplane get away from them. They were warned to keep the attitude of the nose where it was supposed to be during the climb, turns, level flight, and final approach, plus spot checking the airspeed and altimeter.

The next circuit of the pattern was made with the airplane very nose-heavy, and this time I kept my hand in front of the stick to be sure that we didn't dive into the ground. Once again the students were cautioned to keep the airplane's nose in the proper attitude by flying the airplane and not letting it fly them. Flying the pattern twice in these out-of-trim conditions usually gave the students a tired or sore arm, but at least I didn't have to worry about them getting in trouble later with an elevator trim tab (stabilizer) stall.

Go-around on final approach: When the gear and flaps are down and the airplane is trimmed for final approach, even experienced pilots often display sloppy technique when they find it necessary to execute a go-around.

Pilots wanting to make a transition from a descent to a climb add full power for a go-around, but fail to realize the approach trim will cause the nose to pitch upward too much. They intend to establish a climb, yet the airplane immediately enters a climb by itself when full power is applied. Once again this is a case of the airplane flying the pilot instead of the pilot flying the airplane. As power is added for a go-around with the gear and flaps extended, the nose on most airplanes needs to be rather low to maintain airspeed until the gear is raised and the flaps are milked up.

While these tasks are being accomplished, the pilot must overpower the elevator trim with forward pressure until the trim is readjusted. This is a simple fact that most pilots are aware of, yet 75 percent of the people I have flown with over the years perform go-arounds by establishing a climbing attitude before gaining speed and retracting the gear and flaps. A number of airplanes will climb very slowly or not at all using full power when the flaps are fully extended. A Cessna 150 is a good example of this because when full flaps are used on this airplane they are very effective, and there is virtually no climb performance.

Low altitude pull-ups, zooms, buzzing: Although some low altitude steep pull-up stall/spin accidents may result from a high speed stall or accelerated stall, I don't believe that it happens this way very often. Pilots who have a sudden urge to show off are usually the ones who get involved in stall/spin accidents resulting from buzzing. It is my opinion that accidents of this type are more likely to occur as pointed out in the following sample case.

In order to attract the attention of a person or group of people on the ground, the pilot flies low in the vicinity of the spot he plans to buzz. Approaching this spot, a shallow dive is entered to gain airspeed for the pull-up. As the airplane reaches a speed well above the cruising speed for which the airplane was trimmed, considerable forward pressure is necessary to stay in the dive. During the pull-up the airplane is very buoyant because the elevator trim is more effective at this higher speed. The airplane is light on the elevator control; its ability to climb seems effortless and the airplane has come alive. As soon as the pilot opens the throttle the elevator trim becomes more effective and the performance more spectacular. At this point the airplane's nose is above the horizon in a 45 degree pitch attitude and getting steeper. It got into this attitude with little effort on the part of the pilot because of light elevator pressure. The pilot is elated and amazed at the airplane's performance—it's going up like a homesick angel. Soon there is a rapid decay in airspeed because of the steep climb, and more likely no right rudder pressure is being used to correct for torque during the pull-up. Suddenly, the airplane whip stalls abruptly and falls off to the left. What does the pilot do now? He recognizes the fact that this is a stall and proceeds to put into practice what he has learned about stall recoveries by relaxing the back pressure, using full throttle (but it's *already* open), applying right rudder and full right aileron. The ailerons don't seem to be functioning as advertised—maybe they came unhooked. In fact, *none* of the controls seem to be helping the matter. By now the nose is down and starting to rotate to the left, so the pilot continues to hold right rudder and right aileron and changes to full up elevator in an effort to raise the nose. Oh yes, the throttle is still wide open. You know the rest of this story (Fig. 4-6). Oh well, pobody's nerfect. This concludes my series of stories about airplanes flying the pilot.

Fig. 4-6. The dire results of a pilot's first (unintentional) spin. Training to recognize and recover from spins should be an important part of a pilot's training.

THE PIVOTAL ALTITUDE MAKES A DIFFERENCE

Why are more pilots of small airplanes involved with spin-in accidents than pilots of larger, faster airplanes, while attempting to circle around a friend's home? Pilots of slow, easy-to-fly airplanes are accustomed to making small-radius turns as the pivotal altitude of these airplanes is quite low (usually around 500 feet AGL). Too often they attempt tight steep turns at a low altitude. Combined with a misuse of the controls, the airplane stalls and spins. The airplane that cruises at 200 knots has a much higher pivotal altitude, so the pilots flying them do not develop a habit of attempting to circle an object at 500 feet. In the fast airplane, if they *do* attempt to enter a turn at a low altitude around an object on the ground, the object soon disappears from the pilot's view behind the airplane.

The higher pivotal altitude of the faster airplane is also a deterrent to inadvertent spins out of the bottom of the final approach turn. Pilots of fast airplanes start their entry to the final turn sooner. They don't wait until they are nearly opposite the runway and then rack the airplane into a steep bank to line up with it.

GOOD DUAL INSTRUCTION IS A MUST

To show why pilots should never attempt solo spins or acrobatic maneuvers without first receiving dual instruction on these maneuvers, I will describe several "How it should *not* be done" examples from my past experience.

When I learned to fly in 1933, it took one year of washing airplanes to earn flying time on five different types of airplanes (Nicholas-Beasley, Moth, Waco, Command-Aire, and Curtiss Robin) to solo. My last instructor was a private pilot who owned the Challenger Robin (Fig. 4-7) and I was his first student. I can't recall ever making a right turn in the Robin before solo because we remained in the traffic pattern practicing takeoffs, four left turns, hit-and-bounce landings, and go-arounds.

Whenever we were able to make a full-stop landing, the instructor cussed me out real real good as we taxied back for another takeoff. He was in the back seat of the Robin with dual controls but there were no brake pedals back there, so all he could was ride out the swerves on landings and yell at me. Other than a steady stream of obscene and profane language, the only thing he ever told me about flying was, "You are going to tear the (*expletive deleted*) thing up; let's try it again!" After lining up for takeoff and opening the throttle, we would go charging down the field as though I had a wildcat by the tail. When my instructor decided that I was ready to solo, he had another pilot (who held a Transport License) ride with me. Transport License privileges at that time were similar to present day Commercial Pilot Certificate privileges. Since Flight Instructor ratings were not in existence then, holders of a Transport License were permitted to give flight instruction. The transport pilot rode around the pattern with me twice, and then I soloed with a total of 6 hours 41 minutes dual.

A few months later, after checking out in an E-2 Taylor Cub and

Fig. 4-7. Curtiss Robin, 185-hp Challenger similar to the one in which author soloed. Curtiss Robins held the world's endurance records on two occasions, and Douglas "Wrong-Way" Corrigan flew to Ireland in a J6-5 Robin.

hearing my friends talking about doing spins, I decided that it was time for me to try some of them. One of my private pilot friends briefed me on entering and recovering from spins, so I felt qualified but needed to practice them. After climbing the E-2 Cub to altitude and clearing the area below, I applied carburetor heat, closed the throttle, stalled it, and kicked full rudder. As the nose dropped, I chickened out, relaxed the back pressure and returned to level flight. While circling around for a while and telling myself that I was yellow if I didn't spin the Cub, I once again applied pro-spin controls. The second entry attempt produced a nice spin followed by an immediate response to the controls on recovery. It was great—nothing to it! After practicing three more spins, I returned to the airport grinning like a jackass eating briars.

The old E-2 Cub was a good sanitary airplane with no weird spin characteristics. It would recover from a spin immediately by merely releasing the controls. The bad thing though about doing solo spins without any dual spin instruction was the fact that something could have blown off the airplane by gaining excessive speed in the recovery dive. The best advice that anyone might give pilots today is: *"Don't practice solo spins or acrobatic maneuvers without proper dual instruction and supervision."*

Bismarck, a 19th century German statesman, once said, "Only fools say they learn from experience; I have always contrived to learn from the experience of others."

AN UNINTENTIONAL SPIN

Bill Catlin, president of Catlin Aviation in Oklahoma City, tells of an experience one of his salesmen had during the late 1960s while demonstrating a Cherokee 180. The salesman, a 250-hour private pilot, had no previous spin training or spin experience prior to this flight.

While a prospective customer was flying the four-place Cherokee, the salesman was in the right seat and a friend of the customer was occupying a rear seat. They were at altitude when the customer stalled the airplane; suddenly it got away from him and entered a spin. The salesman thought the spin appeared to be going flat, so he turned around and grabbed the passenger by the hair and pulled him forward, nearly to the windshield.

The salesman then tried everything he could think of including opening the throttle in an effort to stop the Cherokee from spinning. Finally, the rotation did stop and the salesman recovered from the spin. It was the first spin that he had seen from the air, and getting the weight up front helped in making the recovery successful.

Sometime after the first Cherokee 180s were built and certificated for four persons in Normal Category, several loaded ones were involved in flat spin accidents. The late Francis R. Keen, a research pilot at the FAA Aeronautical Center in Oklahoma City, was assigned the task of conducting more spin testing on this model airplane. During these tests a Cherokee 180 was loaded with ballast that had been strapped in the rear seats to simulate aft CG loading of passengers. Fortunately, the test aircraft was equipped with a spin chute, because it was necessary to deploy the chute during the tests to recover from a fully developed spin of several turns duration.

When the Cherokee 180s first came on the market I flew them quite often, but most of the time there was only one other person riding with me. I was impressed with the cruising speed of this airplane with four persons aboard when compared to its performance with two people aboard. It seemed to be a better balanced airplane with four people in it. I presume the extra weight in the back seats permitted the stabilator to operate in a more streamlined position, thus reducing the drag and increasing the speed.

Chapter 5

Intentional Spins

A spin is an aggravated stall that results in autorotation and causes an aircraft to describe a downward corkscrew path. Autorotation results from unequal lift or angle of attack of the wings, and the airplane is forced downward by gravity, rolling and yawing in a spiral path.

In the early days of aviation, a spin was called a *tailspin* and usually ended in a fatality. Aerodynamic theory had not advanced to the point of being able to explain a spin, analyze its causes, or determine the proper recovery techniques. Finally, it was discovered that this occurred in the region of reversed command, and the pilot's recovery procedures needed to be opposite from what was considered normal.

When the airplane starts to spin, the difference between the angle of attack of the two wings is considerable. As the airplane descends, the angle of attack of the inner wing is greater than that of the outer wing, causing the inner wing to be more completely stalled. The point here is that the inside and outside wings are both stalled beyond the stall angle of attack and are experiencing different angles of attack. The inside wing will have the higher angle of attack and therefore more drag than the outside wing. This higher drag starts the airplane rotating about the vertical axis and causes the aircraft to continue to spin until the pilot executes a recovery. This difference in drag between the inside and outside wings causes autorotation. From this we can see how lowering an aileron on a stalled wing is similar to having a flap down on a stalled wingtip. Instead of raising the wing, the lowered aileron makes the stall *worse* or converts it into a spin.

While spinning, an airplane descends vertically (because of gravity) and at the same time rotates about a vertical spin axis. This axis is perpendicular to the surface of the earth. While spinning, the axes of the aircraft are all inclined toward this vertical spin axis. Thus, a spin is composed of pitching, rolling, and yawing. Although many conditions may lead to a spin, there are only two factors that actually cause any airplane to spin. These two factors are stall and yaw (*turn*). If either of these factors is not present, an airplane will not spin.

To spin, an airplane must be in an aerodynamic stall. The only thing that causes the stall to occur is an excessive angle of attack. A pilot, then, must first avoid the stall. If this is accomplished, a spin cannot develop. Warnings of the approach to the spin entry depend on the particular airplane. Aerodyanmic buffet, lateral instability, and autorotation are the most common spin warnings an aircraft provides.

Buffet is the turbulent airflow that is generated when the boundary layer separates near the wing root, then passes over the horizontal tail assembly and creates buffet on the elevators. The airplane may shake like a car on a washboard road and this buffet can be felt on the control yoke. There is some turbulent flow generated before the stall actually occurs, so the buffet can occur before the airplane actually stalls and can be used as a warning to recover before a spin develops.

Lateral Instability is an inherent stall warning in swept-wing airplanes. Instead of stalling at the wing root first, swept-wing airplanes stall at the tips first because of spanwise flow. When this happens, tip stalling rarely occurs evenly.

Autorotation simply means that the airplane is automatically rolling and yawing without any input of the pilot. It might result from an external wind gust, uneven fuel balance in the wings, or other factors. There are certain forces acting to keep the airplane in a spin. The effect of these forces is called autorotation. Autorotation is that characteristic of an airplane in a spin which causes it to continue to spin.

Today the average private or commercial pilot has had no spin training. In its place he harbors fear, anxiety, dread, fright, alarm, or trepidation about spins, because he is scared spitless of them. After receiving a license, many pilots even quit practicing stalls because they fear the unknown and believe they might be flirting with a spin.

It would be interesting to know how many present-day instructors have ever practiced solo spins. Some instructors are reluctant to permit students to go very far in performing stalls, which reflects their real fear—spins. This fear, misinformation, and lack of knowledge about spins is then passed on to students.

Not long ago I was talking with a young instructor who told me that he has given 600 hours dual instruction to students in one of the popular new trainers that is certified for spins. He said that he has never given spin training to any of his students, and in fact has never spun this airplane. When asked if he had received any spin training, he answered that he was given dual instruction on spins in two other type aircraft trainers. Yet he had never performed a spin solo.

Prior to 1950, pilots thought no more of doing solo spins than they did of practicing steep turns, stalls, chandelles, lazy eights, etc. A spin was just another maneuver to them, a part of flight training and they accepted it. As kids hanging around an airport, my friends and I were always bumming airplane rides. Some of the pilots we rode with didn't know how to perform many maneuvers other than stalls. They seemed to enjoy watching us hanging on to our seats as they did a series of steep nose-high stalls with steep dive recoveries. The sensation in my stomach during those stall episodes was very unpleasant, and I often wondered if I would ever become accustomed to that uncomfortable feeling in the pit of my stomach. After learning to fly, I soon found these unpleasant sensations completely disappear when you are doing the flying yourself. To those persons who have never been in a spin, I might add that the sensation of a spin is no more unpleasant than doing stalls or steep turns. I thoroughly enjoy doing aerobatics in an airplane, yet to this day I don't enjoy riding on a roller coaster, because being violently jerked around feels unsafe to me. On a roller coaster I feel helpless while experiencing many uncoordinated forces over which I have no control. This may be a result of never receiving proper roller coaster indoctrination.

In a normal spin with the throttle closed, there is no more stress and strain on an airplane than there would be on a leaf falling from a tree, rotating as it falls. With proper training, there is absolutely nothing unduly hazardous in spinning an airplane that is certificated for spins. A spin is no more dangerous than a steep turn.

An average student pilot can quickly learn spin recoveries because they are done by formula rather than by feel. A special control technique, simple but different, is necessary. A spin recovery is one maneuver in flying that is almost entirely *mechanical*. You learn a given sequence of control *movements* and that's it. Memorize the proper entry and recovery procedures for intentional spins and you will know exactly what to do and when to do it.

The spin can be considered to consist of three phases: the incipient spin, steady state (developed spin), and the recovery (Fig. 5-1).

1. The *incipient phase* occurs from the time the airplane stalls and rotation starts until the spin axis becomes vertical or nearly vertical. During this time the airplane flight path is changing from horizontal to vertical, and the spin rotation is increasing from zero to the fully developed spin rate.

2. In the *steady state* (developed spin) the attitude, angles, and motions of the airplane are somewhat repetitive and stabilized from turn to turn, and the flight path is approximately vertical. The spin is maintained by a balance between the aerodynamic and inertial forces and moments in a predictable pattern of motion.

3. The *recovery phase* begins when antispin control inputs are applied and ends when level flight is attained. The specific control movements required in any particular airplane depend on certain mass and aerodynamic characteristics, and are very similar for most small airplanes.

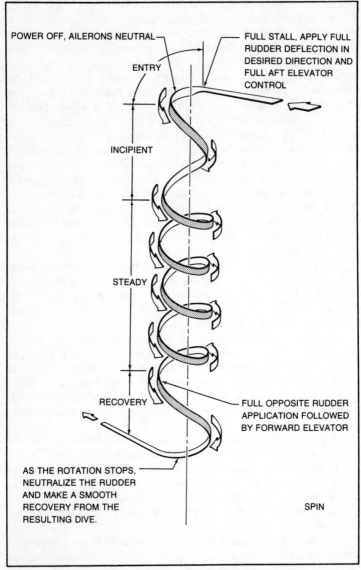

POWER OFF, AILERONS NEUTRAL

ENTRY

FULL STALL, APPLY FULL
RUDDER DEFLECTION IN
DESIRED DIRECTION AND
FULL AFT ELEVATOR
CONTROL

INCIPIENT

STEADY

RECOVERY

FULL OPPOSITE RUDDER
APPLICATION FOLLOWED
BY FORWARD ELEVATOR

AS THE ROTATION STOPS,
NEUTRALIZE THE RUDDER
AND MAKE A SMOOTH
RECOVERY FROM THE
RESULTING DIVE.

SPIN

Fig. 5-1. Anatomy of a spin. (courtesy Cessna Aircraft Co.)

AIRCRAFT MANUFACTURERS' SPIN PROCEDURES

The following list summarizes important safety points relative to the performance of intentional spins in Cessna models approved for intentional spins. This excerpt is from the Cessna booklet titled *Spin Characteristics of Cessna Models 150, A150, 152, A152 (Fig. 5-2), 172, R172* and *177*. Quote:

Basic Guidelines for Intentional Spins

" 1. Know your aircraft thoroughly.

" 2. Prior to doing spins in any model aircraft, obtain thorough instruction in spins from an instructor fully qualified and current in spinning *that model.*

" 3. Be familiar with the parachute, airspace, and weather requirements of FAR 91.15 and 91.71 as they affect your flight.

" 4. Check the aircraft weight and balance to be sure you are within the approved envelope for spins.

" 5. Secure or remove all loose cockpit equipment prior to takeoff.

" 6. Be sure the area to be used is suitable for spins and is clear of other traffic.

" 7. Enter each spin at a high altitude. Plan recoveries to be completed *well above* the minimum legal altitude of 1500 feet above the surface.

" 8. Conduct all entries in accordance with the procedures recommended by the manufacturer.

" 9. Limit yourself to two-turn spins until completely familiar with the characteristics of your airplane.

"10. Use the following recovery procedurs for the Cessna models 150, A150, 152, A152, 172, R172, and 177:

"a. Verify that ailerons are neutral and throttle is in idle position.

"b. Apply and *hold* full rudder opposite to the direction or rotation.

"c. Just *after* the rudder reaches the stop, move the control wheel *briskly* forward far enough to break the stall. Full down elevator may

Fig. 5-2. 1981 Cessna Aerobat certified in Acrobatic Category. (courtesy Cessna Aircraft Co.)

Fig. 5-3. Beechcraft Skipper 77 certificated in Utility Category for spins. (courtesy Beech Aircraft Corp.)

be required at aft center of gravity loadings in some airplane models to assure optimum recoveries.

"d. *Hold* these control inputs until rotation stops. Premature relaxation of the control inputs may extend the recovery.

"e. As the rotation stops, neutralize rudder and make a smooth recovery from the resulting dive."

Each airplane has its own spin behavior and spin recovery characteristics, and pilots flying different makes and models should be thoroughly familiar with the procedures recommended in the Pilot's Operating Handbook or Flight Manual for the airplane being flown. Aircraft manufacturers' recomendations 'concerning spins in other popular present-day training airplanes are included, in order for you to compare them. These excerpts are from the Pilot's Operating Handbook for the Beechcraft Skipper 77 (Fig. 5-3), which is certificated in Utility Category. Quote:

Spins

"The airplane will not spin if orthodox entry is used, but will enter a spiral dive. Speed builds up rapidly in a spiral dive requiring high pullout loads; therefore, if a spiral is inadvertently entered, recovery from the spiral is to be initiated within two turns.

Entry

"Stall the airplane with the control column hard back, throttle in idle position, flaps up, carburetor heat as required and with the nose about 15 degrees above the horizon. At the stall, apply full rudder in the direction required to spin. A slight rudder application immediately before the stall will assure the direction of spin. The airplane nose will drop and rotate towards the applied rudder. When the wings are 90 degrees to the horizon, apply full aileron against (i.e., against the intended direction of spin). The airplane will go slightly inverted and enter a normal spin.

"If aileron against [*the spin direction*] is not applied or applied too late, the airplane will enter a rapid spiral dive, and recovery must be initiated by the second turn.

"If the full back control column is not applied and held, the airplane may spiral. Again recovery must be initiated not later than the second turn. If aileron is applied too early, the airplane will not rotate and merely remain in a straight stalled condition.

Recovery

"Recover from the spin by immediately moving the control column full forward and simultaneously applying full rudder opposite to the direction of the spin. Continue to hold this control position until rotation stops and then neutralize all controls and execute a smooth pullout. Ailerons should be neutral and throttle in idle position at all times during recovery.

WARNING
"Intentional spins prohibited with flaps extended."

Here are excerpts from Piper Aircraft Corporation's Owner's Handbook on the PA38-112 Tomahawk, which is certificated in Utility Category (Fig. 5-4). The Tomahawk is currently the only Piper aircraft being produced that is certified for intentional spins. Quote:

4.41 Maneuvers

"The airplane is approved for certain aerobatic maneuvers, provided it

Fig. 5-4. Piper Tomahawk certificated in Utility Category for spins. (courtesy Louisville Flying Service)

is loaded within the approved weight and center of gravity limits. The approved maneuvers are spins, steep turns, lazy eights, and chandelles.

"Intentional spins are prohibited in the normal category airplane. Lazy eights and chandelles may be performed in the normal category provided a 60 degree angle of bank and/or a 30 degree angle of pitch is not exceeded.

4.43 Spins

"The airplane is approved for intentional spinning when the flaps are fully retracted.

Before Spinning

"Carrying baggage during the spin is prohibited and the pilot should make sure that all loose items in the cockpit are removed or securely stowed including the second pilot's seat belts if the aircraft is flown solo. Seat belts and shoulder harnesses should be fastened securely and the seat belts adjusted first to hold the occupants firmly into the seats before the shoulder harness is tightened. With the seat belts and shoulder harnesses tight check that the position of the pilots' seats allow full rudder travels to be obtained and both full back and full forward control wheel movements. Finally check that the seats are securely locked in position. Spins should only be started at altitudes high enough to recover fully by at least 4000 feet AGL, so as to provide an adequate margin of safety. A one-turn spin, properly executed, will require 1000 to 1500 feet to complete and a six-turn spin will require 2500 to 3000 feet to complete. The airplane should be trimmed in a power-off glide at approximately 75 knots before entering the stall prior to spinning. This trim airspeed assists in achieving a good balance between airspeed and G loads in the recovery dive.

Spin Entry

"The spin should be entered from a power-off glide by reducing speed at about 1 kt/sec until the airplane stalls. Apply full aft control wheel and full rudder in the desired spin direction. This control configuration with the throttle closed should be held throughout the spin. The ailerons must remain neutral throughout the spin and recovery, since aileron application may alter the spin characteristics to the degree that the spin is broken prematurely or that recovery is delayed.

Spin Recovery

"a. Apply and maintain full rudder opposite the direction of rotation.

"b. As the rudder hits the stop, rapidly move the control wheel full forward and be ready to relax the forward pressure as the stall is broken.

"c. As rotation stops, centralize the rudder and smoothly recover from the dive.

"Normal recoveries may take up to 1-½ turns when proper technique is used; improper technique can increase the turns to recover and the resulting altitude loss.

Further Advice on Spinning: Spin Entry

"Application of full aft control wheel and full rudder before the airplane stalls is not recommended as it results in large changes in pitch attitude during entry and the first turn of the spin. Consequently, the initial 2-3 turns of the spin can be more oscillatory than when the spin is entered at the stall.

Spin Recovery

"The recommended procedure has been designed to minimize turns and height loss during recovery. If a modified recovery is employed (during which a pause of about one second—equivalent to about one half turn of the spin—is introduced between the rudder reaching the stop and moving the control column forward) spin recovery will be achieved with equal certainty. However, the time taken for recovery will be delayed by the length of the pause, with corresponding increase in the height lost.

"In all spin recoveries the control column should be moved forward briskly, continuing to the forward stop if necessary. This is vitally important because the steep spin attitude may inhibit pilots from moving the control column forward positively.

"The immediate effect of applying normal recovery controls may be an appreciable steepening of the nose-down attitude and an increase in rate of spin rotation. This characteristic indicates that the aircraft is recovering from the spin and it is essential to maintain full anti-spin rudder and to continue to move the control wheel forward and maintain it fully forward until the spin stops. The airplane will recover from any point in a spin in not more than one and one half additional turns after normal application of controls.

Mishandled Recovery

"The airplane will recover from mishandled spin entries or recoveries provided the recommended spin recovery procedure is folowed. Improper application of recovery controls can increase the number of turns to recover and the resulting altitude loss.

"Delay of more than about 1½ turns before moving the control wheel forward may result in the aircraft suddenly entering a very fast, steep spin mode which could disorient a pilot. Recovery will be achieved by briskly moving the control wheel fully forward and holding it there while maintaining full recovery rudder.

"If such a spin mode is encountered, the increased rate of rotation may result in the recovery taking more turns than usual after the control column has been moved fully forward.

"In certain cases, the steep, fast spin mode can develop into a spiral dive in which the rapid rotation continues, but indicated airspeed increases slowly. It is important to recognize this condition. The aircraft is no longer autorotating in a spin and the pilot must be ready to centralize the rudder so as to ensure that airspeed does not exceed 103 kt (V_a) with full rudder applied.

Dive Out

"In most cases, spin recovery will occur before the control wheel reaches the fully forward position. The aircraft pitches nose-down quickly when the elevator takes effect and, depending on the control column position, it may be necessary to move the column partially back almost immediately to avoid an unnecessarily steep nose down attitude, possible negative G forces, and excessive loss of altitude.

"Because the aircraft recovers from a spin in quite a steep nose-down attitude, speed builds up quickly in the dive out. The rudder should be centralized as soon as the spin stops. Delay in centralizing the rudder may result in yaw and 'fishtailing." If the rudder is not centralized it would be possible to exceed the maximum maneuver speed (V_a) of 103 kt with the surface fully deflected.

Engine

"Normally the engine will continue to run during a spin, sometimes very slowly. If the engine stops, take normal spin recovery action, during which the propeller will probably windmill and restart the engine. If it does not, setup a glide at 75 kt and restart using the starter motor."

STUDENT PILOT DILEMMA

An experienced instructor friend of mine named Floyd Wheeler tells the following story concerning an experience one of his students had with spin recoveries. (Later in the book I will tell you more about Floyd.) Here is one of the spin stories he sent me:

"I remember one of my students telling me that he had been practicing solo spins in a Stearman and nearly failed to recover from a right spin. I couldn't quite understand this, but the next time I rode with him I found out why he was having difficulty.

"I am sure you have noticed that the flatter an airplane spins, the slower it rotates (because of the great amount of fuselage drag), whereas an airplane that spins in a near-vertical attitude has less fuselage drag and consequently a higher rate of rotation. In most airplanes that can be fully stalled, rotation usually continues for ¼ to ¾ of a turn after opposite rudder and forward recovery elevator controls have been applied. As the down elevator is applied, the nose is lowered and the rate of rotation increases momentarily, due to decreased fuselage drag, until sufficient airspeed is attained to break the stall and stop rotation.

"On the next flight with this student, I asked for a two-turn spin to the right because I wanted to observe his entry and recovery technique. After two turns of a right spin, the student applied left rudder and forward stick, then the increase in rate of rotation confused him—and thinking that he had erred, he immediately switched back to right rudder and back pressure. In the near-vertical nose-down attitude we really began to wind up, so he switched again to left rudder and forward stick which stopped the spin rotation. Once he understood to expect that momentary increase in the rate of rotation when anti-spin controls were applied, he had no further trouble."

This is a good example of how a pilot might become anxious after applying spin recovery controls and finds this action seems to be making conditions *worse*—faster rotation and the nose in a lower attitude. In this situation, an inexperienced pilot may become impatient and discontinue the use of anti-spin controls before the airplane responds to the control inputs used for recovery. Pilots are accustomed to having the controls respond immediately when using them during the performance of most maneuvers. But, in spin recoveries they must learn to be patient and hold anti-spin controls long enough to get results. In some airplanes, it takes time for the new position of the controls to produce effective aerodynamic forces.

The preceding incident related by Floyd Wheeler reminds me of an accident which occurred at a primary flying school during World War II, when a cadet in a Stearman PT-17 attempted solo spins for the first time. The cadet failed to release the back stick pressure in the spin and attempted to recover by applying only opposite rudder, which reversed the direction of rotation numerous times before the airplane finally struck the ground. The cadet was only bruised up with no broken bones because the Stearman hit the ground in a flat attitude while making the last change in the direction of spin rotation. This appears to be a case of a cadet who was rather weak military pilot material—or he had a *mighty* weak flight instructor who cleared him for solo spin practice without adequate spin training. The correct answer is probably "all of the above."

PROLONGED SPINS

While attending the 1981 Bowman Air Fair in Louisville, I observed one of the participants in the show perform a prolonged spin demonstration in a modified 1946 Aeronca (Fig. 5-5). The program listed the title of the act as "Tailspin from 6000 feet and deadstick safety demonstration by Bob Livingston." After spinning down to about 1500 feet, Bob Livingston performed other maneuvers (deadstick) before landing and rolling to a stop in front of the announcer's stand. When he was back on the ground I had a chance to talk with him, and asked him to write up some of his experiences with spins or unusual spin stories and send them to me. I am certainly *not* recommending that you strap yourself in an airplane to try the things this man tells about, but he mentions several things concerning spins that most pilots probably haven't considered before. Here are some of the things he wrote about:

"My flying experience spans 32 years as an airport line boy, student, charter pilot, flight instructor, duster pilot, corporate pilot, airshow promoter, and performer. I have spun almost every plane I qualified in, which includes the old Aztec that I flew for three years in corporate work. The

Fig. 5-5. Bob Livingston's 1946 Aeronca 7AC-DC, which was converted from 65 hp to 85 hp and dorsal fin added.

application of ailerons actually worsened the spin and seemed to encourage a flattening of the spin in the instance of the Aztec. I was flying solo and wearing a chute, but the only time I ever thought I might have to get out and float down was when the Aztec let me know it cared not a bit for aileron activity while spinning.

"Why did I spin all these airplanes? Because I wanted to know how each plane reacted under *all* conditions, and I was interested to see how they would react under the 'worst' condition. So I spun 'em. After that, I was more at ease with each plane because we had been 'through it' together. I was free of unnecessary apprehensions and relaxed to fully experience the sensations of flight.

"Unusual experiences? They were *all* unusual; no two spun alike, not even those of the same species spun the same. If someone asked me, 'How does that plane spin?' I would suggest we go spin it and see, or that they go spin it themselves and see. That accomplished two things: It saved me breath and separated the manure from the straw. I have spun many types of aircraft but have never found any to be more pleasant to spin that the small-finned Aeronca 7AC. In my opinion, it is the very best aircraft to teach spins in, and second to that would be the Citabria line. They recover quickly with no nonsense as long as they are well within the weight and balance envelope. The worst airplane I have ever spun was a Tiger Moth in Spain. It had been rebuilt by sloppy workers at a small airport near Zaragoza.

"At a small field one day, a discussion arose over whether an Ercoupe (Fig. 5-6) would spin in spite of all that we had heard about it being 'spin-proof,' etc. This was during what I call my *er* days, when I was young*er*, skinn*er* and no doubt craz*ier*! Embroiled in this hangar debate, I did my usual thing and volunteered to prove an Ercoupe could be spun! I felt there wasn't any plane that would not do a spin. That is, if they had two wings—one on each side—they would spin. The 'experts' surrounding me thought I was a perfect fool to make such a foolish statement. Now committed to follow through, I got in the nearby Ercoupe and took off. A few minutes later the experts were gazing up where I circled at 6000 overhead, announcing via UNICOM I was about to spin down. (See Fig. 5-7).

"Twice I tried the ordinary entry to spin but she would not—not enough elevator. So I dived sharply and pulled up with all the up elevator I could get. Still no spin, only a lot of slopping around in very mild wallows. This was getting serious! All my buddies were now watching me, the greatest pilot around, and I was not living up to my boasts. Determined to spin, I pushed her over and got her going as fast as she would go. I refused to even look at the airspeed indicator, afraid of what it might be registering! I hauled back on the yoke and up we went, then over on our back looking like a loop for sure would result.

"The only thing is, when I got her inverted, she just somehow quit flying. Like dive brakes had been activated, she seemed to just stop all forward movement, and then began to settle inverted. The engine promptly coughed and windmilled, then stopped altogether. I clearly remember saying aloud 'Oooohhh [*expletive deleted*]' with the mike keyed as I fell against the closed canopy in the zero G condition. There were no negative forces—just no forces at all. The plane went totally out of control and I could do nothing at all. I shoved the controls all over the cockpit, but for naught. I forced my eyes to the altimeter and could see it was slowly unwinding, at about a 500 fpm rate. I was helpless, falling upside down, and I remember thinking what a dumb way this would be to die.

"To this day I don't know how, but the plane lurched and fell over right side up into level flight, then rotated once to the left in a sort of flat spin. The motor was still, of course, but at least I was right side up and in a straight

Fig. 5-6. Ercoupe of the type Bob Livingston tried to spin.

Fig. 5-7. Duane Cole down low in his Taylorcraft. Note heavy-duty wing struts. (courtesy Duane Cole)

glide. The altimeter read 2700 feet, so I made the decision to try an airstart. I pushed the plane's nose down and again we dived. When the airspeed passed through 100 the prop reluctantly began to turn, then she started. I landed forthwith and rolled up to where my friends were. Shakily, I extracted myself from the plane and sat down on the grass, struggling to appear nonchalant. One of the fellow pilots announced he had never seen anything quite like that, and wished I would do it again sometime when he had a chance to tell others to come see the demonstration. As gracefully as I could, I declined. I have never tried that move again—and *won't!*

"Note: The reason I had to make an airstart was because the starter had been acting up and was unreliable. As I said, I was in my *'er'* years and determined to fly, regardless of aircraft condition.

"For many years I have done prolonged spins as part of my airshow act, as long as it did not conflict with other performers' plans. The highest altitude I have taken my Aeronca to start a spin down was 10,000 feet. Regardless of the number of turns, she has never even slightly hesitated on recovery. Actually, my plane does not really care to begin a spin. It has the big vertical fin that someone once determined was essential if more than a 65-hp engine was hung on the nose. Later on it was proven the fin made no difference at all, but I like mine and have resisted removing it the several times I've had the tail bones naked. I have to get the nose very high and use a dash of power to help her begin autorotation. When she spins, she does so nicely and there is no stress whatsoever during the spin; only on recovery does she grunt a little. Nowadays, I find that a 20-turn spin is sufficient to please the crowd and fellow pilots.

"I might say another thing about prolonged spins that are done on purpose. The greatest hazard, even for a veteran pilot or performer, is the almost stupefying hypnotic effect they have, particularly when one has been in the spin for a long time. Once the spin is in progress, no strain or discomfort is felt; in fact, a great peace may sometimes come over the pilot. One could almost take out a magazine, light up a cigarette and maybe even take a little nap while spinning down.

"I lost a good friend in Spain that way, I believe. We had often talked about the rapture that was felt while in long spins and chuckled in agreement that it would never affect us because we were too cautious. Yet, while performing near Barcelona one day, he spun down in what was supposed to be just a 10-turn spin, but ended up doing 17 turns into the side of a hill. I was watching through binoculars and never saw him make any effort to recover. I believe he was in fact in the 'spell' of that spin. I do all I can to tell pilots about this when I find they have interest in long spin-downs. Every time I perform my long spin-down I level off first, recall my friend who died spinning, think about my family, the spectators, and the pilots and would-be pilots who are watching me. I speak to my plane and know it will take care of me if I take care of it.

"I have never experienced a full-blown 'flat' spin. For that matter, I never care to, I reckon. No plane I have ever flown *within the envelope* has ever misbehaved. In fact, I find hard evidence to support my belief that all ordinary aircraft must be *forced* to spin, either by deliberate action or gross ignorance enhanced by simple pride.

"I feel strongly that spins should be taught to all student pilots prior to solo flight, along about the seventh hour of training. This I have done with nearly all my students over the years. There are those who say spins are to be considered and taught as advanced maneuvers, but I strongly disagree with that position. Spins are as basic to flight as stalls, steep turns, and spot landings. To say that spins are to be considered in the same category as advanced maneuvers, and that a student should be allowed to fly an airplane without the complete knowledge of how to protect himself from such a potentially deadly flight characteristic, is, in my estimation, near criminal.

"I have a small collection of military aviation training books of the World War II era, and I am satisfied that what students were taught then was correct in every respect. The pioneers of flight—at least those who survived—knew what they were doing because they lived through proving the truths of flight. Only later, unfortunately, did the Feds and other misled sources of influence begin overall tinkering with the training of pilots."

The Ercoupe episode mentioned by Bob Livingston should bring back memories to older pilots of the days when they too were young and cocky. Young people don't always carefully consider the consequences and pitfalls of such an undertaking. If he had been successful in getting the Ercoupe to spin, the question is: Would the Ercoupe have had enough down elevator travel to recover? The Ercoupe had no rudder pedals, but instead incorporated a bellcrank system that hooked the ailerons and small twin rudders together. When the yoke (wheel) was turned to the left or right, both the ailerons and small twin rudders were simultaneously displaced. If this airplane had entered a spin, would there have been enough rudder surface or sufficient rudder travel to stop the rotation? Finally, suppose the Ercoupe was in a vertical dive after stopping spin rotation. The restricted up elevator travel with the yoke all the way back may have been ineffective in preventing a dangerously high airspeed and structural failure of the wings.

DUANE COLE ON SPINS

With his brothers Marion and Lester, Duane organized the Cole Brothers Air Show in 1946. The show was the most outstanding of its kind for 17 years (Figs. 5-7, 5-8). It was terminated with the death of his son Rolly in 1963. Duane won the United States Aerobatic Championship in Phoenix, Arizona, in 1962 and again at Reno, Nevada, in 1964. He has been a major force in the promotion of air racing, and served as Executive Director of the Reno Air Races from 1966 through 1968. He has instructed more hours of aerobatics than anyone in the world. His students have come from as far away as France, Australia, Japan, and Brazil. Still active in the airshow circuit, he has traveled to all 48 of the continental United States to perform, as well as in Canada, Mexico, and behind the Iron Curtain.

Duane is the author of seven books and he has given me permission to quote excerpts on stalls and spins from several of his books. His book *Roll Around A Point* states:

"I know that you have been a little apprehensive about spins, but you will be over your uneasiness as soon as you do one.

Fig. 5-8. Duane Cole beside his clipped-wing T'craft, in which he won the U.S. Aerobatic Championship in 1962 and 1964. (courtesy Duane Cole)

"Although the stall/spin accident has been the greatest killer of all in general aviation, the spin itself is not dangerous. A spinning airplane is a free-falling object rotating about its vertical axis and is subject to no undue strain. The danger of a spin comes from the ignorance of it."

Here is an excerpt from his latest book *Happy Flying, Safely:*

"Statistics prove that a climbing turn improperly executed at low altitude can be dangerous because of the possibility of the stall/spin. To set the record straight, neither the stall nor the spin in themselves are dangerous. Actually, a spinning airplane is a free falling object subject to only one G, the same as imposed on an airplane in straight and level flight. The danger lies in the recovery or lack of it. So, number one in importance is the pilot's ability to recognize a spin, and number two is his ability to stop it. Stopping a spin is a relatively simple thing to do. For instance, to recover from a spin out of a climbing turn, close the throttle, apply full rudder in the direction opposite to the spin and pop the stick forward. When the rotation stops, neutralize the rudder. The airplane will then be pointed down and beginning to accelerate. At this point ease on back pressure to bring the nose up to straight and level, then add sufficient power to keep it there.

"The dangers in the recovery are allowing the airplane to gain excessive speed and loss of altitude before back pressure is applied to complete the recovery, or the application of elevator too abruptly, thus inducing a secondary stall and an additional loss of altitude."

The reason I chose to quote the preceeding paragraph is because Duane is the first author that I know of who has mentioned "pop the stick forward" since that term was downgraded many years ago. It was after I had written my comments on this subject in an earlier chapter that I found this phraseology in Duane's latest book. Finally, in a book he wrote for aerobatic instructors titled *Conquest Of Lines And Symmetry* he has this to say about spins. Quote:

"There has been a bugaboo about spins for as long as I can remember. It started in the old days when the stability of airplanes left something to be desired. But now in this era of stable airplanes, there is no need to fear spins—only the lack of knowledge of them. Teaching spins may be the most worthy accomplishment of an aerobatic course. The ability to recognize and to recover from a spin could someday prevent your student from becoming a statistic. It is my understanding that the stall/spin accident is still aviation's number two killer.

"The spin is being used more and more in National and International Aerobatic Competition so it should be treated as a true aerobatic maneuver rather than an accidental happening. Though your student may have had spins shown to him some time in the past, introduce them as if he had never been in one. Start with one turn but don't do it yourself—let him do it. When he discovers how easily he slipped in and out of the maneuver without undue stress on himself or the aircraft, he will have gone a long way toward allaying any fears he may have had."

Chapter 6

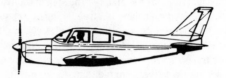

Loss of Control
and Recovery Technique

It is interesting to note that most flight training manuals and the pilot's operating handbooks for airplanes certificated for spins stress that the ailerons be neutral during entry and recovery of intentional spins. As an example of how pilot training in stall recoveries interacts negatively with intentional spin training, consider the following:

Assume the wings of an airplane are completely stalled, one wing drops, and the nose pitches downward as the airplane enters an unintentional spin. Pilots react by doing what they were taught concerning stall recoveries—that is, coordinated use of all flight controls: opposite rudder and aileron and perhaps more relaxation of back pressure. So the stall recovery often becomes an unintentional spin entry (a spin with ailerons opposite direction of rotation). I would be willing to bet money, marbles, or chalk that in recent years many light twins that hit the ground in a flat spin were spinning in the direction of a fully lowered aileron, while partial or full up elevator was also being used.

In my younger days, while experimenting with several airplanes that had been tested and found free of any undesirable spin characteristics, I noted these traits in normal spins. After the spin rotation was stabilized using full rudder and up elevator travel, the application of full aileron in the direction of the spin increased the rate of rotation and caused the nose to lower. When the ailerons were fully deflected opposite the direction of spin rotation, the nose of these airplanes had a tendency to pitch up into a flatter attitude, and the rate of rotation slowed. I hasten to add that it would not be a safe practice to attempt this in many present-day airplanes. This is not meant to imply that all airplanes will react the same way when ailerons are applied in the direction of rotation or against the direction of rotation in a spin. Each airplane has its own spin characteristics, and these characteristics also vary with changes in weight distribution for a particular airplane.

GEOMETRIC TWIST OF A WING

Most modern aircraft are so constructed that the wing will stall progressively outward from the wing root to the wingtip. One method of causing the root to stall first is geometric twist which is nothing more than building a twisted wing. The root section angle of incidence is greater than the tip section. This twist is approximately three (3) degrees and is called *washout*. The washout causes the tip section to have a smaller angle of attack than the root section (Fig. 6-1). Assume an airfoil section has a stalling angle of attack of 18 degrees. When the root section is at 18 degrees angle of attack, it is stalled. However, the tip section is still at about 15 degrees angle of attack and is not stalled. The pilot will feel an aerodynamic buffet from the turbulent air at the root section which serves as a stall warning.

In the example above, the tips have an angle of attack of 15 degrees and are not stalled. At this point, large deflections of the ailerons will cancel out the washout effectiveness and cause the tip section to stall. A lowered aileron would soon exceed three degrees of wingtip washout. In fact, the drag of a full-down aileron at the stall will act as lowered flap on that wingtip.

What pilots have been taught for over 30 years is again quoted from AC 61-21: "The use of ailerons in stall recoveries was at one time considered hazardous due to the inefficient design of some older airplanes. In modern type certificated airplanes the normal use of ailerons will not have a

Fig. 6-1. 1940 photo of 2nd Lt. Everett Gentry (author) in front cockpit of Army Air Corps Reserve BT-9. Note wing leading edge "eyebrows" added to improve wingtip stall characteristics.

detrimental effect in a stall recovery." The word *normal* has been interpreted by most pilots to mean: Use the ailerons at all times, even if the wings are stalled.

In the last ten years, several other writers have spoken out on the pitfalls of using ailerons when the wings are stalled.

It is my belief that the way some pilots use (or misuse) the controls to recover from stalls is the very thing which causes them to enter inadvertent spins. Pilots who persist in using ailerons to keep from entering an upright spin often force the airplane into a flat spin. When the wings are stalled and the airplane does not respond to the use of a little aileron, some pilots continue using more aileron until these surfaces are fully deflected. This is a sure way to stall the airplane's wingtips, which usually causes it to fall off in the direction of a fully lowered aileron. As the nose drops, a pilot without spin training habitually reacts by applying full up elevator in an effort to raise the nose. So down they go with the yoke in a rear corner, and starting with an upright spin, the airplane is soon forced into a flat spin with ailerons. Most airplanes that enter flat spins do not recover from them before they hit the ground.

Prior to the publication of CAA TM 100, all civilian pilots in the U.S. were taught that "approaches to and recoveries from stalls are done with the elevators and rudder only, while keeping the ailerons in the neutral position." Thousands of World War II pilots were also trained in a similar manner. Even today, glider and sailplane pilots throughout the world are being taught this method—that normal coordination of rudder and ailerons should be used in recovery to level the wings *only* after flying speed is regained.

Any modern airplane will respond perfectly in stall recoveries to prompt brisk forward pressure, rudder pressure to keep the nose straight ahead, and ailerons neutral. After the elevators have unstalled the wings, apply coordinated rudder and ailerons to level the wings. Stall recoveries executed in this manner can be accomplished just as quickly as when ailerons are used throughout the stall. It is always the safe way.

Concerning the use of ailerons in stall recoveries, the USAF Primary Flying Student Textbook used in Beech T-34 training states: "If the wings were completely stalled, the use of aileron would be ineffective and aggravate the stalled condition regardless of the finesse with which aileron was applied. The correct recovery technique would be to initiate the proper stick action to break the stall and maintain direction with rudder until the aircraft begins 'flying,' then use aileron to level the wings."

ZERO GS

It seems there is always a hotshot in every crowd and sometimes you may get involved with one without knowing it. This happened to me on one occasion, when another flight instructor and I were flying together on a proficiency flight in a T-6. During the flight we were supposed to change seats and alternate flying instruments under a hood. After I practiced a few

Fig. 6-2. North American AT-6A. This was an instrument trainer at Bryan AFB, Texas, in 1944. The one mentioned in the story was a 1951 re-manufactured T-6G for the Korean Conflict.

minutes of basic instruments under the hood, the other pilot took control of the airplane to give me some "unusual attitude" recovery practice. The gyros were caged and I was flying on partial panel during the recoveries.

Prior to this flight, I had heard several other instructors on the base mentioning that they had performed loops in the T-6 (Fig. 6-2) using a low entry airspeed and dumping flaps in the pull-up. When this guy in the front seat started a pull-up at 130 mph, I assumed we were entering a loop. During the pull-up he dumped the flaps, got the airplane started over on its back, and yelled, "You got it."

When I glanced at the airspeed it was indicating 35 or 40 and decreasing, so I centered the needle and ball and eased forward on the stick. The engine sputtered and continued to windmill (which is what most *radial* engines without inverted fuel systems usually do—the prop keeps turning). By easing forward on the stick just enough to go to zero Gs the airplane acted very well. On the other hand, if I had applied very much back stick pressure or more forward pressure this would have produced a stall. We were on our back settling some as the nose (heavy end) of the airplane started dropping. Gradually, the airspeed began to build up, at which time I retracted the flaps and closed the throttle. From there on it was just a matter of keeping the needle and ball centered in the recovery dive back to level flight.

The point I am trying to make here is that this airplane was flown at 30 mph or lower without experiencing an aerodynamic stall, when normally it would stall somewhere around 60-70 mph power-on or power-off with the wings level. The second point I would like to make is the fact that this airplane would have stalled if the stick had been pushed full forward while inverted. And finally, I would like for you to consider how "going to zero Gs" might be a lifesaver in other inadvertent stall situations.

Incidentally, this pilot killed himself in an airplane about a year after this incident. I heard it was a pilot error accident.

STALLS WITH THE STICK FORWARD

It is possible to stall an airplane at any speed, even at top speed, by simply pulling the control stick back far enough or too abruptly. It is also possible to stall an airplane by pushing the stick all the way forward. Let's consider how several acrobatic maneuvers are *entered* by stalling an airplane with the stick forward.

Inverted spin: One way to enter an inverted spin is to roll to an inverted position using a half slow roll, then close the throttle and use forward stick pressure to hold the nose above the horizon. As the airplane loses speed the stick is pushed full forward until a stall occurs. Just before the airplane stalls, apply and hold full rudder in the direction you desire to spin inverted. To remain in an inverted spin the stick must also be held full forward. This is usually a rather uncomfortable maneuver since the pilot is hanging on the safety belt and being thrown outward from the seat.

Another way that an inverted spin might be entered is during the recovery from a normal upright spin by misuse of the controls. When full rudder and full down elevator are used to recover from a normal spin and the rotation has stopped, it is possible in some airplanes to enter an inverted spin if these control inputs are not released. In other words, if the full down elevator is held too long it causes the nose to tuck under partially inverted and the airplane stalls. If spin recovery rudder is also being held, the airplane will enter an inverted spin in the direction of the rudder deflection.

I remember when the first Air Force T-37 jet trainers arrived at Moore Air Base in Mission, Texas, we were all given thorough spin indoctrination on this airplane because its spin characteristics varied a great deal with various amounts of fuel in the fuselage tank and two wing tanks. A fuel imbalance of 70 pounds in the wing tanks may cause the effectiveness of recovery procedures to vary. A light weight T-37 would enter a spin most quickly. Since much of the fuel is contained in the wing tanks, much of the weight is far away from the spin axis. Therefore, a heavier T-37 aircraft tends to enter a spin more slowly. After spin rotation has stopped in the T-37 and during the recovery dive, placing the elevators in a position *forward of neutral* could result in a transition to an inverted spin.

Outside snap roll: This is the last maneuver I will mention that is also entered by stalling an airplane with the stick fully forward. It is usually more violent than an inverted spin because you are hanging on the belt and the airplane seems to be trying to throw you out of it. This maneuver can be entered from a climbing attitude at less than cruising speed, or it can be entered from an inverted position. In either case, the stick is snapped full forward to produce an accelerated stall and simultaneously full rudder is applied in the desired direction along with full aileron. Performing an outside snap in a Stearman I found works best if the entry is made similar to starting a slow roll. As the bank approaches 45 degrees for a snap to the left, simultaneously apply full left rudder and full left aileron and snap the stick fully forward (all three controls will be in the left front corner and good things should be happening). Note: I have heard that some airplanes perform this maneuver best when the ailerons are crossed on entry.

MORE ABOUT STALLS

It isn't necessary for the airplane to have a relatively low airspeed in order for it to stall. An airplane can be stalled at any airspeed. All that is necessary is to exceed the critical angle of attack, which can be done at any airspeed if the pilot applies abrupt or excessive back pressure on the elevator control.

Airplanes are stalled with the elevator control (or stabilator) and they are also unstalled with the same control. Power is used to assist in the recovery from certain stalls, making it possible to recover without a great loss of altitude. Some pilots believe that opening the throttle alone at the stall will get the airplane under control or they think it is always possible to fly out of a stall with power. The truth is that if the elevator control or stick is held fixed or back when power is added, the stall becomes more vicious. Forward stick movement must be used to unstall the wings by reducing the angle of attack, then the use of power will assist in recovering with a small loss of altitude.

It is not necessary for the airplane to have a relatively high pitch attitude in order for it to stall. An airplane can be stalled in any attitude. Repeating the statement made previously, all that is necessary is to exceed the critical angle of attack. This can occur in any attitude by application of abrupt or excessive back pressure on the elevator control.

When I was instructing students in Aeronca Champions and later in Piper PA-11s, I showed them a spin demonstration to prove that it was possible to enter spins at high airspeeds from nose-low attitudes. These airplanes had a cruising speed of 80-85 mph, yet we entered spins in them at 90-95 mph without placing any undue stress on the airplane. This was done by obtaining 90-95 mph in a shallow dive with power, then the throttle was closed and the airplane smoothly rolled into a 45 degree banked gliding turn (usually a left turn). A gradual buildup of back stick pressure and more bottom rudder pressure was continuously applied until the turn had progressed to the 180 degree point. By this time the airspeed had decayed considerably, so full bottom rudder was applied and simultaneously the stick was brought back in a rear corner (right). Both airplanes entered spins without hesitation. This demonstration convinced students that in order to stall or spin, it is *not* necessary to be flying slowly with a nose-high pitch attitude.

An airplane will stall at the same angle of attack whether with power on or power off. The pitch attitude, however, is considerably different. When practicing stalls, the normal power-on stall and resultant thrust requires a greater pitch attitude than the power-off stall.

Turbulence can cause a large increase in stalling speed. Encountering an upward vertical gust causes an abrupt change in relative wind. This produces an equally abrupt increase in the angle of attack, which could result in a stall. This fact is important to the pilot in that when making an approach under turbulent conditions, a higher-than-normal approach speed should be maintained. Also, in moderate or greater turbulence, an airplane should not be flown above maneuvering speed. At the same time, it should not be flown too far below maneuvering speed, since a sudden severe vertical gust may cause an inadvertent stall due to the higher angle of attack at which it will already be flying.

When Stearman PT-17s were built for the Air Force, spoilers were

Fig. 6-3. Stearman PT-17s at Mississippi Institute of Aeronautics, Madison, Miss. The PT-17 was one of the best trainers ever built.

installed on the entire length of the leading edges of both top and bottom wings. This was done to make the Stearman fly like a heavier and faster airplane. Abrupt application of back pressure on the elevator control would cause these airplanes to shudder and stall. Stearmans built as Navy trainers had the spoilers installed only on the leading edges of the lower wing panels (Fig. 6-3).

HIGH SPEED STALLS

A high speed stall or accelerated stall might occur during the recovery dive from a spin or the last part of a loop or similar maneuver. This happens when the pilot applies too much back pressure and exceeds the critical angle of attack, which causes the airplane to buffet or shudder. Normally, the recovery from a high speed stall can be accomplished by merely releasing some of the back pressure. Since the airspeed is high, releasing some or all of the back pressure produces an immediate recovery. All a pilot needs to do is to get the wing lined up with the flight path of the airplane—or stated another way, align and keep the wing chord in line with the direction of relative wind. If the airplane has started to roll or snap in a high speed stall, it may be necessary to use a more positive recovery with some down-elevator deflection to produce an immediate recovery. Note in Fig. 6-4 how airplane B was in a high speed stall; however, when the chord line of the wing was realigned with the relative wind, the wing was unstalled (airplane position C).

When an airplane is in a steep dive after recovering from a spin, the pilot must start applying back stick pressure immediately to keep from gaining excessive airspeed. The higher the airspeed, the greater the G forces will be during pull-out. If abrupt or excessive back pressure is applied during pull-out this may produce an accelerated stall. It must be remembered that when an airplane is headed straight down, the pilot must take prompt action to start it on its way back to level flight before the airspeed runs wild. This is not a good time to stop and smell the roses.

What factors cause an increase in load factor? Any maneuvering of the airplane that produces an increase in centrifugal force will cause an increase in load factor. Turning the airplane or pulling out of a dive are examples of maneuvering that will increase the centrifugal force, and thus produce an increase in load factor. When you have a combination of turning and pulling out of a dive (such as recovering from a diving spiral) you are, in effect, placing yourself in double jeopardy. This is why you must avoid high speed diving spirals—if you accidentally get into one, be careful how you recover. Turbulence can also produce large load factors. This is why an airplane should be slowed to maneuvering speed or below when encountering moderate or greater turbulence.

AILERON REVERSAL AND WING DIVERGENCE

There are many clean, fast airplanes being flown by general aviation pilots these days; for this reason, several other critical factor topics are

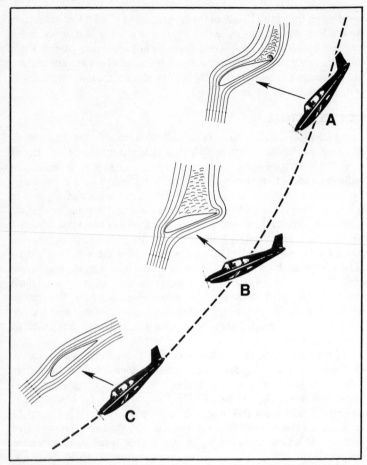

Fig. 6-4. High-speed or accelerated stall. An airplane can be stalled in any attitude at any airspeed.

included in this book for your consideration. The U.S. Naval Aviation Safety Center published a booklet called *Flyboy* which contained reprints of articles from their *Approach* magazine files. Here are two exerpts from a 1955 Approach article entitled "Something's Gotta Give." Quote:

Aileron Reversal

"This matter of aileron reversal is a study in contradictions, as anyone who ever experienced it can tell you. Say you're tooling along, in the redline zone, with just a normal 1G showing, mind you, and you shove the stick over to the left for a port turn. So what does the airplane do but roll itself into a *right* turn, leaving you just the slightest bit confused about these mulish machines. Now don't hit the panic button yet; just throttle back, pop some speed brakes (extend some garbage, as some put it) and let's look at what happened.

"You were nudging the redline speed on your cloudhopper, remember? Now, as you're boring through the blue, with no aileron control applied at all, your wing is subject to a down pitching moment if the airfoil has any camber. But just you roll in some aileron and the pitching moment (or torsion) can be up *or* down due to aileron displacement.

"It works this way: Roll in a bit of *down* aileron on one wing at redline speed and a considerable force immediately acts against the aileron to force it and the adjacent wing area *upward*. This results in a twisting effect on the wing which bends the leading edge *down*, and vice versa for the *up* aileron.

"In times past, airplane designers have been able to beef up the wing structure to take care of most any twisting effect the wing might encounter. However, as speeds increased, it was obvious that the only way to prevent this twisting due to the pitching moment of the wing from deflected aileron, particularly near the wingtip, would be to build the wing like an anvil. But if you build an airplane like an anvil, it'll fly like an anvil.

"So, in spite of a very fine strength factor built into our new airplanes, there still remains the twisting moment which tends to bend the leading edge of the wingtip up or down according to aileron deflection.

"At redline speed, straight and level so far as ailerons are concerned, the wing at this point is about ready to toss in the towel. If at this time you roll in left stick for a left bank, the right aileron dropping into the airstream causes the right wing to groan 'that does it!' and the right wingtip is twisted down to give a lift down instead of up.

"In other words, it gives a lift change contrary to what we are controlled for. You see, we put our right aileron down, that causes the wing to twist leading edge down to give a lift down on that side. It also works the reverse way on the left wing. Left aileron there tends to twist the wing leading edge up and give us a lift up instead of down. Result: A perplexed pilot sitting there with a mess of ineffective aileron on his hands. Remedy: Come back outta that redline zone before you reach the original point of no return. All this is a high indicated airspeed problem, and the approach to aileron reversal may give some warning, but not very much. For the record, swept-wing aircraft are more susceptible to aileron reversal than straight wing aircraft."

Wing Divergence

"Wing divergence—now there's a real nasty sounding hunk of terminology, and believe you us, it's even nastier than it sounds. This caper can put you in the past tense before you can say 'next of kin.'

"Observe: Wing divergence occurs because a given wing can take just so much stress acting upon it in this manner. There's upward stress, from the lift; there's backward stress that tends to pull the wings backwards, and both of these are pretty well licked by design. Then there's torsional, tip-bending stress that tends to twist the wingtips. This twist causes them to dig in and this dig-in bends the wings up or down. That's right, up or down—and *all the way*. Wing divergence, friends, is strictly a one-shot deal—there's no vibration, no buffet, no shudder, it just starts bending and twisting and the wings are gone.

"Again, let's say you've gone messing around in that redline zone, sort of moth-to-the-candle-like. The process of wing divergence can be seen like so: If a wing at high indicated airspeed encounters a slight updraft due to turbulence, there is generated a slight uplift which causes a momentary increase of lift; next, this momentary increase in lift causes the wing to deflect upward giving further increased angle of attack; this increased angle of attack gives more lift, which is more deflection, which is more angle of attack, which is more lift, until failure, when the wing diverges.

"That means the wing went wham! bam! and was no longer a part of the airframe. The wings just went up and back like a swan diver's arms. But don't bother to begin worrying now—you just became 'was' from the shock of the divergence.

"You T-28 and AD drivers should know that wing divergence is most common to straightwing airplanes. Swept-back wings have a stabilizing effect for this situation. (You get more of the problem of aileron reversal in the swept-wing type.)

"If by now you are still unimpressed by the nature of the *things* which lurk in the redline area, be advised that your curious or heedless probing into this region can flip the lid on a Pandora's Box of aeroelasticity and compressibility troubles which could flip you smack into the obituary column.

"So, to avoid any possibility of *that* conclusion, consider this conclusion: Recognize the limitations of your aircraft; respect the redline speeds assigned by the people who built the machine; operate within the airplane's performance envelope and in accordance with the pilot's handbook and you've got it locked. Horse around in the marginal areas and brother, you never had it so wild!"

WINGOVER WITH PASSENGERS

After obtaining a Limited Commercial license in 1936, I started hopping passengers in my friend's model R Stinson (Fig. 6-5). This was a four-place cabin monoplane powered with a 225-hp Lycoming engine. It was a rugged, fine airplane and as stable as a Brunswick pool table.

One day, three of my high school buddies asked me to take them for a ride in the Stinson. Within a few minutes, they were bored with level flight and turns and yelled for me to do some stunts. After doing several steep turns and stalls, they still wanted something more exciting, even though several of them had been hit in the back of their heads with cans of Simoniz and polishing cloths. During stall recoveries, these items came forward from the baggage area behind the rear seats.

The only other maneuver that I could think of was a "wingover," as I had done a few of them in small lightplanes. My first wingover attempt in the Stinson was entered from a shallow dive and a steep pull-up to the left with full power. In a few seconds we were headed straight down with the throttle closed. That heavy old airplane was really building up speed like a streamlined brick, and I was becoming concerned with getting the nose back to level flight without shedding any heat shingles. While applying smooth back pressure—as much as I dared—and feeling like a moron who had reverted back to an idiot, the airspeed continued to build up. The airspeed indicator had numbers from 0-200 mph, and it was reading about 40 on the second time around.

My passengers thought that last maneuver was great and wanted to do it again, but I informed them our time on the flight was all used up. We returned to the airport with me sweating and smelling like a wet hunting dog.

Of course, this incident has nothing to do with stalls or spins, but it does show what can happen with a clean or heavy airplane in a steep nose-down attitude. It is another example of why pilots should have proper instruction before fooling around with an airplane.

There are several other things I learned about flying in the old Stinson, but I will mention just one of them before moving on to another subject. The following episode resulted from immature judgment, cockiness, and over-confidence on my part.

Although the model R Stinson was not equipped with flaps, I soon learned in hopping passengers that it could be slowed down on final approach and landed short. Whenever I was a little too high on final in a power-off glide, I found that by raising the nose some the airplane would sink more rapidly. This flatter approach was maintained until reaching approximately 100-150 feet above the ground, at which time the nose was lowered to pick up more speed. Just before touchdown the airplane was flared with a large increase in the angle of attack (in ground effect) and this

Fig. 6-5. Model R Stinson, originally designed for a retractable gear. This airplane is similar to the one in which Eddie Stinson was killed.

cushion effect provided a smooth three-point landing. There was no floating; the airplane greased on and didn't roll very far after touchdown.

One day while I was making one of these approaches, a friend of mine was sitting on the ground in the shade of an airplane wing talking with several other pilots. This group of pilots included the famous Art Goebel, who was in town doing some skywriting at that time. Art Goebel noticed my landing approach and commented to the group that he was not favorably impressed with it. Later my friend told me what he said, and consequently, that was the last time I used that procedure for landing. I had not given any thought to what might happen if turbulence was encountered while flaring for a landing.

When young pilots are unsupervised, many of them go through phases of cockiness and overconfidence. If they are fortunate enough to scare themselves real good occasionally, it usually serves to straighten them out for a while—that is, until they enter another period of overconfidence. Someone once made this observation: "Confidence is that feeling you have before you really understand the problem, while overconfidence is simply a bomb waiting to explode."

Edward A. "Eddie" Stinson was a civilian flight instructor during World War I at Kelly Field, San Antonio, Texas. He was the first instructor to teach Army pilots spin recoveries in the OX-5 Jenny. In 1918, at Kelly Field, Eddie became world champion looper with 150 consecutive loops in a Thomas-Morse Scout. Some writers claim that Eddie was the first person to proclaim the use of forward stick pressure for recovering from a spin. This may or may not be true, since an Austrailian named Harry Hawker performed the first intentional spin in June of 1914.

Eddie Stinson settled in Detroit, and the first Stinson airplane was built there in January of 1926. His tragic death occurred in January of 1932, while attempting an emergency landing in a Chicago school yard. He was flying a model R Stinson, similar to the one in Fig. 6-5, and was low on fuel. After landing he struck a flagpole and died a few hours later at the age of 38. The company that bore his name continued building these fine airplanes for another twenty years after his death. Starting with the Stinson 105 and later the Voyager, Stinson airplanes were built by Vultee. After Vultee quit building Stinsons, Piper Aircraft acquired the remaining Stinson Station Wagons and parts, which they sold until the supply was depleted.

EARLY SPIN RECOVERIES

The first accounts of unintentional spins that I have found mentioned in aviation history books refer to them as "spiral dives" and "tailspins." A few pilots successfully recovered from spiral dives or tailspins as early as 1911. An Englishman named Lt. Wilfred Park of the Royal Navy, flying an Avro biplane in 1912, discovered that the use of rudder opposite the direction of rotation was the proper technique for recovering from an unintentional spin. Prior to this he had tried to recover using full rudder in the direction of rotation with the control wheel all the way back, plus full power. There was no mention of what he did with the control wheel during his successful recovery.

Harry Hawker, the Australian flying a Sopwith Tabloid biplane in June of 1914, was the first known pilot to attempt an intentional spin, from which he also successfully recovered. The day before, he had attempted a low altitude loop while demonstrating the Tabloid to the Royal Navy in England, and fell into an unintentional spin. He was uninjured when the airplane hit in a clump of trees. Hawker claimed that opposite rudder was stopping the spin and the airplane would have recovered, if he had been higher. The day following his crash, Hawker climbed another Tabloid to 3000 feet and intentionally spun it several turns and recovered. Once again we can wonder what position the control wheel was in during the recovery. Would those airplanes recover by merely relaxing some of the back pressure or did Hawker use forward pressure? Were some of the tailspins referred to at that time actually no more than diving spirals? Was Eddie Stinson the first person to use forward pressure during spin recoveries?

There are a number of small airplanes that will recover from a spin by merely releasing one or both of the controls (rudder and elevator). In fact, it is unwise to use full down-elevator for spin recoveries in Cubs and Aeroncas. Neutralizing the elevators on either of them produces a smooth, prompt recovery, whereas full forward pressure would cause these airplanes to go beyond vertical over on their back. It would be interesting to know if Harry Hawker released some back pressure or if he actually used full down elevator as Eddie Stinson was teaching in the Jennies.

Chapter 7

Stall/Spin
Training in Gliders

Glider pilots learn to fly the glider's wing since that is all they have—there is no engine. For those airplane drivers who believe that power is what they need most to recover from stalls, I want to remind them that power stalls are more vicious than power-off stalls. This is especially true when power is added while the stick is held back. Airplane pilots must also learn to fly the airplane's wing; they must learn how to make changes in the angle of attack so the air flows smoothly over a wing's top surface and is deflected downward.

Someone extolling the excellence of sailplanes made this statement: "An airplane has a noisy fire hazard out front that always sounds like it is trying to destroy itself. This is something you don't have to worry about in a sailplane."

Although neither the demonstrations of spins nor spin training are required by FAA for Private or Commercial Glider Pilot certification, The Soaring Society of America has taken the strongest possible stand urging that all glider pilots receive spin instruction both on the ground and in the air to the point of proficiency in spin entry and recovery. Many glider schools will not permit students to solo until they have received spin training.

While circling in thermals, sailplane pilots operate at very low airspeeds, often nibbling at a stall. If turbulence is encountered while performing these turns, the sailplane will stall at higher-than-normal airspeeds, and misuse of the controls at this point might result in an unintentional spin. Abrupt use of the controls or excessively large deflections of the control surfaces create drag, and each deflected surface acts as a brake which further reduces any remaining airspeed.

Thermaling with a sailplane to gain altitude is accomplished by circling in thermals. This is done at minimum sink speed, which is usually five to seven mph above the stalling speed for the angle of bank being used. Because of the sailplane's long wingspan and normally small radius of turn, when making turns at minimum sink speed it is usually necessary to hold top aileron pressure (opposite the direction of turn) to keep the bank from steepening. The outside wing in a turn travels a further distance than the inside wing and thus generates more lift which tends to steepen the bank. It is also often necessary to maintain some rudder pressure in the direction of turn in order to keep the yaw string or slip-skid ball centered. If they are centered, the turn is correctly coordinated, even though the controls are crossed. Using the rudder and aileron controls in this manner is necessary at low airspeeds. However, when the sailplane is being flown in the traffic pattern at 60 mph prior to landing, the effectiveness of rudder and aileron controls is in better proportion to one another. In other words, these controls can usually be neutralized (the control surfaces are permitted to streamline themselves) after the bank has been established, while turning at higher airspeeds. The yaw string or slip-skid ball will remain centered, indicating a coordinated turn.

Spin training provides the glider pilot with the feel or control touch necessary to recognize when a stall is on the threshold of an unintentional spin. If a student glider pilot falls into an unintentional spin on the first attempt at circling in a thermal, it would be nice for him to know how to recover. Don't you agree?

Portions of the material in this chapter were excerpted from Schweizer Aircraft Corporation's *Soaring School Manual*, and The Soaring Society Of America, Inc. book titled *The Joy Of Soaring*.

HISTORY AND BACKGROUND OF SOARING

Motorless flight as a sport dates back to the very early days of the power plane era. The earlier glider programs conducted by Lilienthal, Chanute and the Wright Brothers, were, for the most part, carried on to develop stability, control techniques and systems for the development of powered aircraft. In 1911, while testing a new control system, Orville Wright made a flight of 9 minutes and 45 seconds. This was the first recorded sustained soaring flight.

Following World War I, interest in the modern soaring movement started in Germany and Europe. In the 1920s, a number of meets were held in Germany, France, and England. It wasn't until 1929 that the sport was first introduced in the United States. The first National Contest was held at Elmira, New York, in 1930.

During the 1930s interest in gliding spread rapidly throughout the country. Considerable numbers of primary gliders were produced, some by established aircraft companies, others by homebuilders in basements and garage workshops. Due to the marginal quality of the equipment and the lack of good instructional facilities and standardized procedures, a number of unfortunate accidents occurred. These incidents had a dampening influence on the sport that lasted for many years.

During the years immediately preceeding World War II, sailplane design improved and an effort was made to standardize training procedures. Two-place trainers were developed as were higher performance single-place sailplanes.

In 1940, the American distance record was 263 miles and the best altitude was 17,264 feet. It is interesting to note that the present distance record is more than three times as far and the altitude record nearly 30,000 feet higher.

Gliders played an important role in World War II. Many of the pilots were trained in Schweizer-built TG-2s and TG-3s. When these ships became available on the surplus market, they made up a large percentage of equipment used by the soaring clubs that became active following the war.

By the end of the 1940s, standardized training programs were being adopted by most of the clubs. At the same time, several full-time commercial soaring schools were established. New developments in sailplane design and construction techniques were introduced in the years following World War II. Performance and safety were combined for the first time. The introduction of the all-metal structure was an important factor in the excellent safety record compiled by these new sailplanes. The theme became, "We do this for fun, so let's do it safely."

During the 1950s, American soaring came into its own. Paul Mac-Cready was the first American to compete in an International Contest. American teams were entered in World Contests held in Spain, England, France (won by MacCready), Poland, Germany and Argentina. This period was marked by the emergence of American soaring pilots as world record holders. Dick Johnson first broke the distance record previously held by the Russians with a flight of 530 miles. In July 1970, Ben Greene and Wally Scott, each flying ASW-12s, soared 717 miles from Odessa, Texas, to northern Nebraska to establish a new world's distance record. In April 1972, Hans-Werner Grosse of West Germany shattered it with a 907 mile flight from near Hamburg, Germany, to the Spanish border. During this period, membership in the Soaring Society of America doubled, then tripled. Many new clubs were formed and more commercial schools were established.

In the mid 1950s, the first true, one-design class sailplane was introduced in America—the Schweizer 1-26. Interest in the class concept continues to grow. Each year a number of regatta-type meets are held in different sections of the country. With all pilots flying identical sailplanes, the emphasis is on piloting skill and ability—competition in its truest form.

General aviation continues to become more demanding of the pilot—efficient operation of equipment increasingly exacting. A growing number of professional and amateur power pilots are taking up soaring because it puts the fun back in flying. Here is flying in the truest sense and finest form.

The challenge it offers can be controlled by the individual. There is a great deal of satisfaction in meeting this challenge, whether the flight is a local one, of an hour duration, or an FAI Award flight for endurance, distance, or altitude. The fascination and satisfaction continue to increase. With experience come increased ability, confidence, and enjoyment.

Today, with soaring centers located in nearly every section of the country, it is possible to take instruction or to soar at a commercial school or with an established club near your home. Some outstanding soaring sites are located in areas that provide an interesting vacation for the entire family. Because of improved equipment, training procedures, and facilities, American soaring has gained national recognition as the ultimate in sport aviation.

SCHWEIZER AIRCRAFT CORPORATION

As schoolboys in 1929, Paul A. Schweizer and his brothers William and Ernie started building sailplanes. In 1936 they started selling them. The present company, Schweizer Aircraft Corporation was formed in 1939, at Elmira, New York.

Schweizer Aircraft is the largest builder of sailplanes in the United States and North America, but unfortunately, their production is still relatively small. Sailplane building is only about ten percent of their business. The rest is doing major subcontract work for the aircraft industry and building the complete AG-Cat agricultural airplane. In 1945, the Schweizer Soaring School was started and their *Soaring School Manual* was developed from the experience they have acquired over the past 37 years; it is updated every few years.

In recent years, with the advent of the commercial soaring school and standardized training program, soaring has gained national recognition. Two factors are primarily responsible for its acceptance as a safe and thoroughly enjoyable sport—the postwar two-place trainer, and standardized training and operating procedures.

The Schweizer 2-33 (Fig. 7-1) is recognized as the best two-place club

Fig. 7-1. Schweizer 2-33—most successful trainer with clubs and schools.

Fig. 7-2. Schweizer 1-36 Sprite—high-performance sailplane for the 1980s. All-metal, Diamond C performance.

trainer in the country. It is used by nearly all of the leading commercial soaring schools and clubs.

Schweizer developed a new all-metal sailplane for the 1980s called the Sprite (Fig. 7-2). It was designed to fill the demand for an affordable sailplane with Diamond C performance, having a 31:1 glide ratio. Since 1939, Schweizer has produced more than 90 percent of all sailplanes made in North America.

AERO TOW AND GROUND LAUNCH

Let's take a brief look at the methods of getting sailplanes into the air.

Aero tow: A tow plane should have good short field capability and be able to climb at speeds from 50-75 mph without overheating. The tow rope is usually about 200 feet in length. As the sailplane is being towed behind an airplane, the climb is quite shallow. The Schweizer 2-33 trainer is usually towed at 60 mph, while the placard speed (never exceed speed) for aero tow is 98 mph. The placard speed for auto or winch tow is 69 mph in the trainer.

Ground launch (auto tow or winch launch): There is very little difference in the technique required between either of these methods of ground launch. Usually, a steel cable is used for the winch launch because it is wound around a drum. If a steel cable 5000 feet in length were used for a winch launch, by the time the sailplane reached its maximum altitude and release point, a considerable amount of the cable would be wound around the winch drum. The release point altitude would be much lower than 5000 feet. The two speed for either ground launch method is rather critical because the maximum allowable speed, surface wind velocity, and other factors must be considered in determining a proper two speed.

The initial climb is made with a pitch angle of about 15 degrees at 50 feet of altitude, and gradually steepened until reaching a maximum of 45 degrees at 200 feet. This is similar to flying a kite, since the stick will be in the full back position with the sailplane's speed *increasing* as the climb is steepened. Pilots are accustomed to applying back pressure and having a reduction in speed; however, in a ground launch the reverse is true. If some back stick pressure is relaxed while climbing, the tension on the tow rope or cable will be less, causing the sailplane's speed to lower. Altitude gained during an auto tow would depend on the length of the runway, the surface wind velocity, length of tow rope, climb angle, etc.. After releasing the tow rope, it may be possible for the pilot to gain more altitude in thermals; otherwise it will be necessary to remain in the traffic pattern for a landing. If a pilot fails to release the two cable during a ground launch and glides beyond the winch or tow car, backward pull on the cable will operate an automatic tow hook release.

STALL TRAINING IN SAILPLANES

Sailplane pilots are given aero tow and ground launch emergency procedure training so they will react properly in case the towline breaks. Since ground launches are performed using a steep climb (45 degree pitch attitude) at low airspeeds of 45-50 mph, prompt recovery action must be taken to keep from stalling after the two cable breaks.

The Joy Of Soaring is a training manual written by Carle Conway for The Soaring Society Of America, Inc. Carle Conway earned his pilot's license in 1928, and has been flying ever since. During World War II, he was an instructor in the USAF and later a Squadron Commander in the Air Transport Command.

Mr. Conway holds a Commercial Pilot Certificate with single and multi-engine, land and sea, instrument and instructor ratings. He has a Commercial Glider Pilot Certificate and instructor rating. He also holds a U.S. Gold Badge with three diamonds.

The following excerpts from *Joy Of Soaring* should show that glider pilot stall recovery procedures are not taught in a timid wishy-washy manner. Quote number one states:

"Stall recovery, through frequent repetition, becomes a 'conditioned reflex' that responds the instant an impending or actual stall is detected. Since the training is directed toward meeting a low-altitude emergency, the recovery should be made to a wings-level, best-glide-speed attitude, with minimum possible loss of altitude. Three vigorous control movements may have to be made at the same time:

1. Stick *fully* forward*, and quickly, but don't leave it there.
2. Spoilers closed.
3. Top rudder pedal fully forward.

*In the typical training glider, 'stick fully forward' is unnecessary because the elevator travel is so restricted that it is difficult to force the wing to a fully-stalled angle of attack. Yet the purpose of stall training is to develop habits that will protect the pilot in the future, when he may be flying gliders with entirely different and more vigorous stall behavior. So 'stick fully forward' should be the way to learn in the trainer, and the instructor should be relied upon to prevent the student from nosing the glider down beyond the angle needed to streamline the wing."

Quote 2: "Top rudder is used whenever one wing is lower than the other at the moment of stalling. In some gliders (not the training type) the use of aileron in a stall to pick up a low wing will have just the opposite effect. The down aileron stalls the low wing even further. At the stall, the rudder is still effective, and if fully applied will help to level the wings and to overcome any tendency of the glider to spin in the direction of the low wing. The instant the elevator unstalls the wing, the ailerons will respond normally, the ailerons and rudder should then be used with normal coordination to level the glider."

Quote 3: "*The fourth stall* simulates conditions when the rope breaks in the steep part of an auto or winch tow. Spoilers are closed for this stall, and the pitch attitude is about 40 degrees nose-up. It is assumed that the pilot was slow in reacting to the break and that a straight ahead stall results. Recovery requires the stick to be *all* the way forward until after the nose crosses the horizon, when it is moved slowly back. The recovery dive should be about the same degree below the horizon as the stall was above it. As speed increases, the nose is brought up smoothly to a normal gliding attitude, being careful not to bring the stick back too far and thus provoke a secondary stall."

SPIN TRAINING IN SAILPLANES

Schweizer sailplanes are noted for their gentle stall/spin characteristics. The entire section of Schweizer's *Soaring School Manual* on spins is quoted for your information and comparison with airplane spin entry and recovery procedures previously mentioned. You will note these procedures are directed to glider flight instructors. Quote:

"*Spins:* Some students will be found to have developed a subconscious aversion to spins and one of the objectives of the instructor should be to eliminate this fear and replace it with respect.

"It has been estimated that there are actually several hundred factors contributory to spins, and from this it is evident that, whether or not spinning is a desirable maneuver or characteristic, it will be a feature of aircraft for some time to come and must be reckoned with in the training of a pilot.

"It is essential that all students be taught to spin and recover before solo. Such knowledge is a great confidence builder, as well as contributing toward development of technique.

"Spin instruction should be divided into two phases: That prior to solo should be confined to teaching the fundamentals of entry and recovery and developing confidence which will enable the student to react properly should he fall into an accidental spin during his early solo work; that given after solo, during the practice of precision maneuvers, will teach the student orientation under adverse conditions, accuracy of judgment, and timing applied to control action. The first phase only will be taken up at this point.

"Logically, demonstration of spins should follow the practice of stalls, but in no case should spin training be undertaken at an altitude of less than 1500 feet, higher if possible, because of the irretrievable loss of altitude in sailplanes.

"To start, the ship should be stalled completely by pulling the stick straight back and holding it firmly in this position. As the sailplane stalls, full rudder should be applied in the direction in which the spin is desired and held there as firmly as possible. *No ailerons should be used.*

"As the nose falls and the ship starts into the spin, the student should have his attention again directed to the sensations he felt just prior to and during the entry. These should be analyzed and cataloged in the memory.

"Care must be taken to completely stall the sailplane, otherwise it may not spin and the only result will be a skidding spiral of increasing speed. The instructor will know the individual characteristics of the ships in which he trains and will be able to impart to his students little knacks which will aid them to make clean spin entries and obviate the possibility of these spiral dives.

"Many sailplanes have to be forced to spin and require considerable judgment and technique to get them started. This judgment and technique are important since they demonstrate the presence or absence in the student of the 'feel' of the spin. Paradoxical as it may seem, these same sailplanes that have to be forced to spin may be accidentally put into a spin by the pilot or student who has failed to develop this feel. The instructor will use his judgment at this stage in applying to his student anything except the basic knowledge of spin entry and recovery.

"*Incipient spins:* Sometimes when a sailplane stalls, one wing may drop and more altitude will be lost than in a wings level stall. This may happen whether in level flight or while turning and is called an *incipient spin* or the first stage of a full spin. If quick corrective action isn't taken, a full spin will develop.

"If the wing drops during a level stall, apply opposite rudder and ease forward on the stick until the nose is well below the horizon. As the sailplane regains flying speed, the wings can be brought back to level. If corrective action is taken promptly, only a slight wing dropping will result with a minimum loss of altitude.

"If the sailplane is stalled while turning, the inner wing will drop. The stalling speed is higher than normal so that in steep turns a quick recovery can be made by just easing forward on the stick. This is because at the higher speed, good control is regained the moment the wing is unstalled.

"Extra care must be taken when turning below 500 feet. A safe margin of speed (depending on how steep the turn is) must be maintained. Because of turbulence and wind gradient effect near the ground, the recovery may take several times as much altitude as a recovery from a stall in good air. Most spin accidents are the incipient ones which happened too near the ground for full recovery. The student pilot should have enough dual practice on the type spin so that he recognizes and takes corrective action immediately.

"Full spins: The entry into a fully developed spin is in two stages. The first or incipient stage is the stalling and dropping of a wing with rotation started. If the sailplane is one that will spin, and if no corrective action is taken, then the second stage or full spin develops. The motion is then automatic, and will continue until a normal recovery is made.

"Recovery from the first or incipient stage is made by simply applying opposite rudder and easing the stick forward. These movements stop the rotation and unstall the wing almost instantly.

"Recovery from the full spin takes longer and much more altitude will be used. First, apply full opposite rudder and after a very slight pause, ease steadily forward on the stick until the spin stops. As the spinning stops, neutralize the rudder and recover to normal flying attitude. It also must be noted that when the spinning stops, the sailplane will be in a very steep dive and that care should be used not to exceed the redline on the pull-out. Use the dive brakes to restrict speed, if necessary, and do not make an abrupt pull-out as this could result in overstressing the sailplane, or an accelerated stall.

"With practice, recoveries from spins can be made with a minimum loss of altitude and at a reasonably slow airspeed. High airspeed is the result of using too much forward stick on the recovery. This, of course, uses up considerably more altitude.

"Although it is important that a student be able to recover from a full spin with a minimum loss of altitude, it is more important that he will not get in an unintentional spin to begin with. Therefore, proper and adequate instruction and practice of incipient spins is most important.

"No properly or adequately trained pilot will ever fall into an accidental stall or spin from any normal maneuver."

SAILPLANE STORIES

While I was with the FAA, three Airman Examination Specialists were assigned the responsibility for developing glider pilot written tests. I was one of them. Each of us was allotted eight hours proficiency time per year in sailplanes, so we usually flew together in an Aztec or Baron from Oklahoma City to Caddo Mills, Texas, in order to rent sailplanes for practice. There is a large sailplane operation at Caddo Mills called Southwest Soaring, Inc., which is located on an old Air Force auxiliary field. This field is about 20 miles from the former Greenville AFB, Texas (a World War II basic flying school). Flying several different model sailplanes only eight hours during the spring and summer months each year didn't mean that we were always proficient—just dangerously current most of the time.

I recall several rather unusual incidents that occurred while flying sailplanes. One day during takeoff, I was on an aero tow behind a Cessna 150 towplane. As the 150 became airborne, its nosewheel dropped down and turned 90 degrees to one side. A bolt in the nosewheel strut scissors had failed, which allowed the strut to extend to its bottom limit and turn sideways. My sailplane was not equipped with a transmitter and receiver, so I was unable to advise the pilot of his nosewheel difficulty. At first, I considered releasing from the towplane and rushing back to the field to alert the pilot of his nosewheel problem before he landed. But I decided that someone on the ground surely noticed the nosewheel position and would advise the pilot of it on UNICOM.

I released from the towplane at 2000 feet and circled over the airport to watch the 150 as it landed. It so happened that no one on the ground observed the nosewheel position, so the pilot was unaware of his predicament. The 150 landed on a concrete runway in a tail-low attitude on the main wheels. As the pilot lowered the nose during the after-landing roll, the nosewheel spun around and then continued to shimmy from side to side until the 150 rolled to a stop. There was no damage to the airplane except perhaps less tread remained on the nosewheel tire. (This story doesn't relate to stalls or spins, but it is nice to know what might happen when a landing is attempted with the nosewheel turned sideways.)

On another occasion during an aero tow behind a Super Cub, I was flying a Schweizer 1-26. After takeoff and upon reaching an altitude of approximately 200 feet, I noticed the towplane rocking around as it flew through a dust devil. Several seconds later, the 1-26 entered the same dust devil which flipped it on its side into a vertical bank. Within a few seconds the 1-26 was out of the turbulence and the wings were level again. However, the towplane was not in sight. I soon discovered the towrope was extending upward from the nose of the sailplane at a 45 degree angle, still attached to the towplane. There was no problem returning to the high tow position (five to ten feet above the towplane's wake), because we were back in smooth air.

After landing and discussing this episode with the pilot of the towplane, he told me that when the Super Cub (Fig. 7-3) entered the dust devil his seat moved rapidly to its rearmost position. The adjustable seat slides in tracks, and since his seat was not properly latched in the ratchet, turbulence caused it to slide to the rear. As a result, he inadvertently jerked the stick back, which placed the airplane in a steep climb.

This could have been disasterous for the towplane, because it was near a stall towing a sailplane which was on its side in a maximum drag position. For a few seconds, this was like dragging an anchor for the towplane and it could have stalled and entered a spin. Without the pilot's seat malfunction, the towplane and 1-26 would have passed through the dust devil with less difficulty.

Large dust devils can create serious situations when they are encountered by aircraft close to the ground. I remember watching an airplane as it got flipped on its back by one. This airplane was landing and about ready to touch down when the right wing entered a dust devil that caused it to cartwheel.

A prominent soaring pilot named Marion C. Cruce participates in soaring meets throughout the country. He lives in Oklahoma city, and for a

Fig. 7-3. 1980 Piper Super Cub similar to towplane mentioned in text. This is a popular towplane aircraft. (courtesy Louisville Flying Service)

number of years he served as a glider instructor and flight examiner. He holds a gold badge with three diamonds. The highest altitude that he has flown in his Schweizer 1-26 is 32,000 feet, although he has been above 30,000 feet in it at least a dozen times. Marion told me of a spectacular accident that he witnessed which involved an aero tow:

"During takeoff, a Super Cub using a 200-foot towrope was towing a Schweizer 2-33. The sailplane was being flown by a student who had made several previous solo flights. While Marion and a friend were assembling a sailplane, someone yelled that a towplane had crashed. There was smoke coming from the towplane which burst into flames after crashing on the airport at a 60-70 degree angle. When Marion looked up and saw the 2-33, it was headed nearly straight up at an altitude of approximately 300 feet. Marion and his friend jumped in a car and rushed to the scene of the crash, expecting the worst. When they got to the towplane the pilot had already managed to crawl out of it. The tow pilot was hospitalized with two broken legs.

"The student in the 2-33 was able to recover from near vertical flight, and landing straight ahead on the airport without scratching the sailplane. When Marion told me this story, I wondered if the student had been recently practicing ground launches with the stick held back on takeoff for a steep climbing attitude. But he assured me this was not the case."

Marion believes that the student got too high on tow and this raised the tail of the towplane, causing it to dive. With the towplane in a dive, the sailplane's nose was being pulled downward, and to compensate the student continued to apply more and more back stick pressure. When the sailplane released from tow—either by the student, a towrope break, or a weak-link (built-in safety factor) break in the towrope—the back stick pressure caused the sailplane to assume near-vertical flight. In a case like this the tow pilot can also release the towrope by use a two-release. This particular towplane was equipped with a two-release that could be activated by kicking it with your heel, as it was mounted on the floor between the rudder pedals."

Marion ended his story by saying, "When I'm flying the towplane and someone starts pulling my tail too high, I release them. If they are not doing what they are supposed to do back there, I get shed of them!"

UNINTENTIONAL SPINS

Robert Gains, Soaring Society of America, former safety chairman, and editor of the "Safety Corner" section of the SSA journal *Soaring*, published an article in the December 1978 issue concerning an unusual close encounter experienced by soaring pilot George Worthington. This Safety Corner article included an introductory statement by Robert Gaines, and is quoted in its entirety:

Thermaling at low altitude is very risky. George Worthington warns that soaring pilots who attempt such maneuvers should be intimately familiar with the stall/spin characteristics of their particular sailplane.— *Robert Gaines, SSA Safety Chairman.*

"I was shocked and saddened to read (*Soaring*, July 1978) of Jim George's death in his AS-W 12. I owned and flew that ship for 700 hours before selling it to Jim about four years ago. I imagine that there are seven or eight AS-W 12s still flying in the U.S. Some of these ships are bound to be sold in coming years to pilots wishing to move up in performance. I can't help thinking of these pilots and wanting to send a little warning their way.

"Jim George was a better pilot than I. He had truly extensive sailplane, and particularly AS-W 12, experience. And yet, judging from the information available, Jim probably stalled and spun while thermaling 'several hundred feet' above the terrain near a gliderport. I don't wish to discuss the risky issue of thermaling so low. Instead, I would like to relate my experience with stall/spins in the AS-W 12. Let me first say that in my opinion, the AS-W 12 has no aerodynamic faults. I don't believe it is more spin-prone than any other high-performance sailplane. I found it to be docile and obedient in the air. However, on about five occasions during my first 300 hours in the 12, I did inadvertently and suddenly enter a spin while thermaling. During the last of the occasions at about 1000 feet above the desert near El Mirage Dry Lake, I was clumsy and heavy-handed in my recovery response. The ship had stalled, nosed down to a 60 degree attitude, and turned a quarter turn. I pushed the stick forward for a fraction of a second and then impatiently brought the stick back too far in an effort to bring the nose up to level flight so that I could continue thermaling.

"As I brought the stick back, G forces increased to more than +1, and I entered an accelerated stall/spin. This time the ship's reaction was more violent. The nose pitched down to 90 degrees. I got the message and was determined not to stall again, so I pushed the stick forward for a definite fraction of a second longer than last time. This pushed the nose down past vertical to about 115 degrees. I knew that I now had to be careful to not tear the wings off as the airspeed was quickly building up to redline. I also knew the ground was closer than I would have wished for. I'm writing this, so you know I made it—by a bare 200 feet.

"I haven't gotten to my main point yet: the foregoing is only an introduction. My main point is that I took that experience rather seriously, and I wondered about the possibility of acting more efficiently should it happen again. I thought about it and decided to try a new response to stall/spins that occur while thermaling. The next time it happened, many hours and weeks later, I somehow knew it was starting and instantly flicked the stick an inch or so forward and just as quickly brought it back to almost its original position. If you have been 20 feet away at that moment, watching the AS-W 12 closely, I believe you would not have been able to detect any nose pitch down at all. The thermaling process was unbroken. During the next 400 hours the ship 'tried' to stall/spin while thermaling on six or seven occasions. Flicking the stick instantly forward always nipped the stall in the bud. I think I flew my last 400 hours in the 12 with an increased degree of alertness and a conditioned response which were sufficient to cope with the problem. I did not change my thermaling technique.

"I've written the foregoing to suggest that the AS-W 12 is near-perfect in the air, but we pilots are not. But even though we are not, there is usually a satisfactory solution to every aerodynamic problem. Is it possible that Jim George might be alive today if we could just have sat down for an hour and discussed the thermaling stall/spin problem in the AS-W 12? Could it help you?"

In the previous article, the writer makes several points which I have tried to stress throughout this book. That is, whenever he was on the threshold of a spin he *instantly flicked* the stick forward which nipped the stall in the bud. Also, he did not leave the stick forward and there was no excessive diving or loss of altitude. The recovery was instantaneous and the thermaling process was unbroken. Some pilots may believe that relaxing the back pressure would have been an effective method of recovery in the AS-W 12, when actually a more aggressive action was necessary to prevent a spin.

Note: Excerpts from *Joy of Soaring* by Carle Conway that appeared in this chapter are courtesy of The Soaring Society of America (SSA). SSA is a division of National Aeronautical Association. The 1-26 Association and The Vintage Sailplane Association are divisions of the SSA. An affiliate of the SSA is the National Soaring Museum. For further information on the sport of soaring, contact the SSA at Box 66071, Los Angeles, California 90066.

Chapter 8

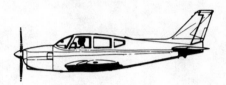

Stalls and
Spins in Light Twins

The stall characteristics of twin-engine airplanes operating with symmetrical power, or with both engines idling, are somewhat similar to the stall characteristics of single-engine airplanes. The major difference between flying a twin-engine airplane and a single-engine airplane is knowing how to properly handle the controls if one engine loses power for any reason (Fig. 8-1). Loss of power from one engine affects both climb performance and controllability of any light twin. Some light twins operating on one engine will maintain level flight with all of the seats occupied, but some of them will not. Many of the old light twins were equipped with engines of low power and when fully loaded would not climb on one engine. These airplanes had marginal single-engine performance capability.

The FAA General Aviation Accident Prevention Program published an excellent handout entitled *Flying Light Twins Safely*. Excerpts from that publication discuss several light twin terms and performance speeds ("V" speeds), and the following information is quoted:

Climb Performance

"Climb performance depends on an excess of power over that required for level flight. Loss of power from one engine obviously represents a 50% loss of power but, in virtually all light twins, climb performance is reduced by at least 80%.

"The amount of power required for level flight depends on how much drag must be 'overcome' to sustain level flight. It's obvious that, if drag is increased because the gear and flaps are down and the prop windmilling, more power will be required. Not so obvious, however, is the fact that drag

Fig. 8-1. 1982 Beechcraft Baron B55 (foreground) and E55 are typical light twins. (courtesy Beech Aircraft Corp.)

also increases as the square of the airspeed while power required to maintain that speed increases as the cube of the airspeed.

"Thus climb performance depends on four factors:

☐ *Airspeed:* Too little or too much will decrease climb performance.

☐ *Drag:* Gear, flaps, cowl flaps, prop and speed.

☐ *Power:* Amount available in excess of that needed for level flight.

☐ *Weight:* Passengers, baggage and fuel load greatly affect climb performance.

Yaw

"Loss of power on one engine also creates yaw due to asymmetrical thrust. Yaw forces must be balanced with the rudder.

Roll

"Loss of power on one engine reduces propwash over the wing. Yaw also affects the lift distribution over the wing causing a roll toward the 'dead' engine. These roll forces may be balanced by banking into the operating engine.

Critical Engine

"The critical engine is that engine whose failure would most adversely affect the performance or handling qualities of the airplane. The critical engine on most U.S. light twins is the left engine, as its failure requires the most rudder force to overcome yaw. At cruise, the thrust line of each engine is through the propeller hub.

100

"But at low airspeeds and at high angles of attack, the effective thrust centerline shifts to the right on each engine because the descending propeller blades produce more thrust than the ascending blades (P-factor). Thus the right engine produces the greatest mechanical yawing moment and requires the most rudder to counterbalance the yaw.

Key Airspeed for Single-Engine Operations

"Airspeed is the key to safe single-engine operations. For most light twins there is an:

	Symbol
Airspeed below which directional control cannot be maintained.	V_{mca}
Airspeed below which an intentional engine cut should never be made.	V_{sse}
Airspeed that will give the best single engine rate-of-climb (or the slowest loss of altitude).	V_{yse}
Airspeed that will give the steepest angle-of-climb with one engine out.	V_{xse}

Minimum Control Speed Airborne (V_{mca})

"V_{mca} is designated by the red radial on the airspeed indicator and indicates the minimum control speed, airborne at sea level. V_{mca} is determined by the manufacturer as the minimum airspeed at which it's possible to recover directional control of the airplane within 20 degrees heading change, and thereafter maintain straight flight, with not more than five degrees of bank if one engine fails suddenly with:

☐ Takeoff power on both engines;
☐ Rearmost allowable center of gravity;
☐ Flaps in takeoff position;
☐ Landing gear retracted;
☐ Propeller windmilling in takeoff pitch configuration (or feathered if automatically featherable).

"However, sudden engine failures rarely occur with all of the factors listed above and, therefore, the actual V_{mca} under any particular situation may be a little slower than the red radial on the airspeed indicator. However, most airplanes will not maintain level flight at speeds at or near V_{mca}. Consequently, it is not advisable to fly at speeds approaching V_{mca} except in training situations or during flight tests.

Intentional One-Engine Inoperative Speed (V_{sse})

"V_{sse} is specified by the airplane manufacturer in new handbooks and is the minimum speed at which to perform intentional engine cuts. Use of V_{sse} is intended to reduce the accident potential from loss of control after engine cuts at or near minimum control speed. V_{sse} demonstrations are necessary in training but should only be made at a safe altitude above the terrain and with the power reduction on one engine made at or above V_{sse}. Power on the operating (good) engine should then be set at the position for maximum continuous operation. Airspeed is reduced slowly (one knot per second) until directional control can no longer be maintained or the first indication of a stall obtained.

"Recovery from flight below V_{mca} is made by reducing power to idle on the operating (good) engine, decreasing the angle of attack by dropping the nose, accelerating through V_{mca}, and then returning power to the operating engine and accelerating to V_{yse}, the blue radial speed.

Best Single-Engine Rate-of-Climb Speed (V_{yse})

"V_{yse} is designated by the blue radial on the airspeed indicator. V_{yse} delivers the greatest gain in altitude in the shortest possible time, and is based on the following criteria:

☐ Critical engine inoperative, and its propeller in the minimum drag position

☐ Operating engine set at not more than maximum continuous power.

☐ Landing gear retracted.

☐ Wing flaps in the most favorable (i.e., best lift/drag ratio) position.

☐ Cowl flaps as required for engine cooling.

☐ Airplane flown at recommended bank angle.

"Drag caused by a windmilling propeller, extended landing gear, or flaps in the landing position will severely degrade or destroy single-engine climb performance. Single-engine climb performance varies widely with type of airplane, weight, temperature, altitude and airplane configuration. The climb gradient (altitude gain or loss per mile) may be marginal—or even negative—under some conditions. Study the Pilot's Operating Handbook for your specific airplane and know what performance to expect with one engine out. Remember, the Federal Aviation Regulations do not require any single-engine climb performance for light twins that weigh 6000 pounds or less and that have a stall speed of 61 knots or less.

Best Single-Engine Angle-of-Climb Airspeed (V_{xse})

"V_{xse} is used only to clear obstructions during initial climbout as it gives the greatest altitude gain per unit of horizontal distance. It provides less engine cooling and requires more rudder control than V_{yse}."

THE FLAT SPIN PROBLEM

Accident reports in recent years show that several of the popular light twins have a propensity for entering flat spins. A number of flat spin accidents occurred with only two persons on board the airplanes, so aft CG loading was not the problem in many cases. If the rear seats had been filled with passengers, the aft CG loading would no doubt have made the airplanes more prone to flat spins. One accident report stated: "Witnesses saw the aircraft spinning with the tail lower than the nose." Most of the light twins were engaged in engine inoperative practice maneuvers involving pilots flying solo or flight instructors on training flights performing V_{mc} demonstrations. Other accidents involved pilots who were faced with actual engine failures or malfunctions. There is a preponderance of evidence that flat spins and doubtful recoveries are related to low speed/high asymmetric power conditions. Recovery from power-off and power-on stalls (both engines) can be safely accomplished in these airplanes using normal stall recovery techniques. Unfortunately, the flat spin accident, so prominent in past years, is still with us.

It is a fact that the ultimate results of a fully developed flat spin are predictable and catastrophic. There have been a few cases where test pilots have been able to recover from a flat spin by "pumping the control yoke" or using unorthodox asymmetrical engine power opposite the direction of spin rotation, but most often a spin chute was deployed to stop the spin.

For a number of years I have flown all of the light twins that have a propensity for flat spins, and considered most of them fine airplanes. In fact, I did not realize that these airplanes were involved with flat spins until reading of the problem in accident reports years later. My flying in the twins included many practice V_{mc} demonstrations, simulated single-engine landings, and other simulated single-engine practice maneuvers, yet at no time was there any evidence of unusual characteristics observed in the aircraft. I am sure the reason there were no uncommon hazardous situations experienced is due to the fact that instantaneous vigorous recovery action was always initiated.

Some flight instructors and pilot examiners seem to take pride in the reputations they have earned of surprising students with simulated single-engine failures under the most unusual circumstances. Often an instructor or examiner may have a well-known inclination to cut one engine at speeds below published V_{mc} to test a student's reactions. Many times these traits are handed down from instructor to instructor, or from examiner to instructor. Pilots who ignore procedures and cautions contained in the pilots' operating handbooks for their aircraft must lack common sense. An official of one of the companies that manufactures light twins stated: "Careful study of the facts of stall/spin accidents in all types of aircraft leads inescapably to the conclusion that the problem is a *people* problem, not an *aircraft* problem." I would like to add that what is being taught as "proper stall recovery procedures" is a large part of the problem.

In a twin-engine airplane, the single-engine power-on stall is characterized by a rapid roll toward the inoperative (dead) engine. If not immediately arrested, this roll progresses rapidly into a wingover or split-S entry into an upright or flat spin. Vigorous and immediate recovery action is required.

Let's consider how a pilot might inadvertently enter an upright or flat spin in a light twin. For instance, imagine a typical multi-engine training flight where the instructor is giving a student some V_{mc} demonstrations. The instructor retards the left throttle to idle while the right throttle is fully advanced. Perhaps the student reacts rather slowly in performing any of the correct engine-out procedures. He will probably be slow in applying adequate right rudder to maintain directional control. When right aileron pressure is applied to bank five degrees into the right (good) engine, more aileron deflection than necessary is often used to cover up for inadequate use of right rudder.

As the airspeed continues to decay, additional right rudder is applied until the pedal reaches the stop. Now, rudder will no longer maintain directional control and the nose swings a few degrees to the left. The student decides it is time to recover so he relaxes the back pressure. To stop the airplane from rolling to the left, more right aileron pressure is added until the ailerons are fully deflected. The airplane stalls with asymmetrical full (right) engine power producing the same effect as the application of full left rudder. Immediately, the nose drops as the airplane rolls on its back and enters an upright or flat spin to the left. If the student fails to close the right throttle and maintains full right aileron, the beginning of an upright spin will most likely be converted into a flat spin within a microsecond.

Pilots of multi-engine airplanes should be thoroughly familiar with the "V" speeds for their airplanes. On the other hand, under certain conditions the single-engine stall speed might be above the V_{mc} speed. In other words, the airplane might stall while operating on one engine before it reaches V_{mc} speed. Rather than relying entirely on numbers on the airspeed indicator when operating at low speeds using one engine, the pilot must be prepared to use proper recovery controls in a prompt and vigorous manner. If the student pilot doing the V_{mc} practice exercise had reacted properly, he could have stopped the light twin "dead in its tracks" from entering a spin . In the first place, the pilot should be aware of how much the ailerons are deflected, using just enough right aileron, but no more than necessary to bank five degrees into the good engine. Directional control should be maintained perfectly straight ahead with right rudder. As soon as the rudder pedal hits the stop, the pilot should realize the situation is reaching a critical point. When the nose of the airplane moves one or two degrees to the left it means that directional control has been lost with the full right rudder deflection. The time for action is *now!* Instantaneously bring both throttles to idle position, and briskly "pop the yoke forward" enough to reach a zero G condition. This would place the airplane in a straight-ahead dive with the nose about 30-40 degrees below the horizon. If the ailerons are fully deflected, remove some of the deflection until the wings are flying. The airplane should be under control immediately and the rudder pedals can be neutralized, since the asymmetrical right engine power was eliminated by closing that throttle. If the right rudder deflection is maintained, the nose will swing to the right as the speed builts up. As soon as V_{mc} speed is attained, the pilot can return to straight-and-level flight using one or both engines. If properly executed, the airplane will remain on the same heading used at the start of the V_{mc} demonstration.

How would you like to ride as a passenger in a 707 or 747 jet airliner if you knew that the pilot planned to operate it at low speeds with all engines on one side idling while those engines on the other side were operating at full thrust? Don't you think you might be a witness to an upright or flat spin? When we ride the airlines we don't worry about such things because we feel the pilots know the limitations of their equipment, its "V" speeds, and characteristics. The professional airline pilots also know their own limitations.

Here is another example of how the pilot of a fully loaded light twin might get into trouble during a single-engine landing, following the loss of one engine. Suppose that this pilot is on final approach with a strong headwind. The gear has been lowered and partial flaps were extended too soon. The pilot realizes that he will undershoot the runway, so full power on the good engine is applied. The critical point can now be quickly reached if the nose is raised to level flight, since the airplane may not be capable of level flight with the drag of the gear and flaps. In this case, the nose pitch attitude must remain low to maintain the desired airspeed and directional control. The pilot must maintain directional control with the rudder.

On the other hand, if the rudder pedal on the side of the active engine is against the stop, this should warn the pilot that loss of directional control could occur at any moment. At this point there are not too many options available. The nose must be lowered to gain speed. A reduction of power on the good engine will provide directional control without the need of full rudder. Milking up the flaps while maintaining airspeed and directional control might save the day. Otherwise, it may be necessary to land straight ahead before reaching the airport. Anyway, if V_{mc} is not maintained throughout the approach, while full power is used on the active engine, the airplane will not make it to the airport.

AEROBATICS WITH FEATHERED PROPS

If you haven't seen R. A. "Bob" Hoover of El Segundo, California, at an airshow performing aerobatics in his Shrike, OV-10, or P-51 (Fig. 8-2), you have missed an exciting experience. With one engine feathered on the Shrike, Bob Hoover performs four, eight and sixteen-point rolls into the dead engine, followed by loops, Immelmanns, and Cuban eights. After feathering both props, his sequence of aerobatic maneuvers includes a zero airspeed stall, sixteen-point barrel roll, eight-point roll, three loops (the last one on the deck), a 180 degree turn, and several sideslips (Fig. 8-3). The landing is then made (with both props feathered), alternating the touchdown on one main gear and then on the other (Hoover's trademark

Fig. 8-2. Bob Hoover and his P-51. (courtesy Rockwell International)

"Tennessee Walk"). Excessive speed in the after-landing roll permits him to coast to a stop and park in front of the speaker's platform (Fig. 8-4).

In response to a question about the advisability of teaching pilots to spin an airplane, Bob said, "I am tickled to death that I was taught spins when I learned to fly. I am sure that it saved the day for me several times in my early years of flying."

Fig. 8-3. Bob Hoover performing aerobatics with both props feathered. (courtesy Rockwell International)

Fig. 8-4. Bob Hoover landing Shrike on one wheel with both props feathered. (courtesy Rockwell International)

Even though some pilots get into trouble with stalls and spins in light twins, Bob performs aerobatic maneuvers with *both* props feathered by converting the high airspeed obtained while recovering from a maneuver into energy, which permits him to perform the next in a sequence of maneuvers.

When asked if he had ever spun the Shrike, Bob stated that he had never made over a half-turn spin in that airplane. He did say that he has talked with several pilots who have spun the Shrike and was told that it spins nicely with no problems on recovery. (However, it would be reasonable to assume that the rear seats were *not* full of people when the spins were performed.)

For many years, expert aerobatic pilots have put on spectacular demonstrations with stock airplanes that were not designed for acrobatic maneuvers. For example, during the 1930s, Harold Johnson was looping a Ford Tri-motor at airshows across the country. This was a six-ton, 14 passenger airliner. Johnson later performed a complete aerobatic routine in a four-engine Lockheed Constellation. It must be remembered that expert aerobatic pilots are capable of flying their airplanes smoothly through acrobatic maneuvers without placing undue stress on the aircraft. They know know to keep from overloading the airplane with G forces. An inexperienced pilot who attempts to show off can easily overstress an airplane by trying to force it through maneuvers with abrupt and mechanical control inputs. Even inputs. Even though the wings of an Acrobatic Category airplane are built to support six Gs (six times the normal weight of the airplane), abrupt and excessive use of up elevator in a high speed dive could cause the wings to separate from the airplane.

In 1954, Bob Hoover was flight testing the North American F-100A near Edwards Air Force Base in California. Wind tunnel work had shown that this airplane might go flat in a spin, from which there is usually no recovery. During the flight tests, he climbed to 42,000 feet and put it into a right spin, using a crossed control technique. Immediately, the F-100 went into a flat spin rotating about one turn per second. He tried all types of recovery techniques, holding each of them four turns, with no results. he then deployed the spin-recovery chute, but it just flopped around and was ineffective because the drag chute lines were too short. When Bob got down to about 15,000 feet he had made 22 turns, so he jettisoned the canopy. With the canopy gone there was no wind in the cockpit since there was no forward motion of the airplane. The centrifugal force was so great that he had difficulty reaching the ejection seat handle. After ejecting at around 7000 feet and free-falling away from the airplane, he opened the chute and the ejection seat fell free. During the free fall he had gone way below the airplane, and as it continued to spin flat, it came very close to him while he was swinging in the parachute. The F-100 hit the ground in such a flat attitude that there were undisturbed bushes standing between the wings and horizontal tail assembly.

After the malfunction of Bob's spin-recovery chute, NASA Langley Research Center improved the design of the chute and determined the correct length for the suspension lines. Suspension lines that are too short cause the chutes to flop around in the turbulent air behind the airplane, yet when they are too long the chutes are ineffective.

MULTI-ENGINE STALL/SPIN STORIES

When small children touch a hot stove and burn their fingers they learn two things: To say the word *hot* and to never again touch a hot stove. If a teenage driver is involved in a fender-bender accident with the family car or gets a ticket for speeding, he becomes more conscientious about his driving habits. With the proper attitude, he develops into a careful driver and is less apt to have a more serious accident in the future. Airplane pilots should learn from the mistakes of others—the price of experience is too high. A pilot certificate should be used as a certificate to *learn more about flying* and to gain more knowledge and skill in the never-ending process of safety education.

Several of my friends have supplied me with these multi-engine stall/spin stories. I believe they have a message for light twin pilots to consider.

An Overloaded Airplane

John Paul Jones, who wrote the article "Minimum Standards and the

Pilot" (in Chapter 2), described the following two accidents which resulted from aft CG loading of aircraft.

The first accident involved a Twin Beech C-45 (surplus military airplane) that cracked up in Alaska sometime during 1960. The pilot of this airplane arrived at an outpost airport to pick up nine carpenters who had been working on a job in the area. Since half of the cabin seats had been removed from the airplane, some of the men sat on the floor. Their winter clothing, baggage, and tools were also placed beside them on the cabin floor. An accident investigation team determined after the accident that the Twin Beech was loaded nine inches outside the aft CG limit and approximately 1000 pounds over certificated maximum gross weight.

The takeoff was made on a long runway at a sea level airport with a temperature of approximately minus 40 degrees Fahrenheit and the very dense air provided good wing lift and propeller efficiency. The high oxygen content of the air resulted in each engine producing its full potential of power. During the takeoff, no difficulty was experienced in raising the tail and the airplane was kept on the runway until it reached 125 mph before liftoff. The excess speed and dense air gave good control on takeoff *while the weight was being carried on the wheels*. After becoming airborne, climb was established and the gear retracted. As the angle of attack was increased for the wings to carry all of the weight, the airspeed decreased. Then it was necessary to apply down elevator to hold the tail up. Because of the rearward CG and additional weight, the pilot ran out of down elevator travel with the yoke full forward. At about 75 feet in the air, the aircraft was approaching a stall and started to fall off on the left wing. This was corrected by a reduction of power on the right engine and use of right rudder, which soon produced a rolling movement to the right. This was corrected in a similar manner—power reduction on the left engine, left rudder, and full power on the right engine, etc. These rolling moments from side to side continued with alternate use of power until the airplane pancaked to the ground and slid about 60 feet. Fortunately, no one was killed in this accident although all the occupants spent time in the hospital with back injuries.

Accident number two concerns a similar situation that happened in Canada after an additional gas tank was installed in a J-3 Cub. The gas tank manufacturer had obtained approval for the modified 10-gallon tank installation for placement behind the rear seat in the baggage area. When a mechanic installed the tank he misread the installation specifications, measuring the distance in inches aft of the datum to the front edge rather than to the center of the tank, which placed the tank too far to the rear.

Two persons were aboard the J-3 on its first flight after modification. The new tank was filled with ten gallons of gas which was to be used on a long cross-country flight. The pilot had no problems on takeoff while the weight was still on the wheels; however, after becoming airborne the airplane climbed steeply to approximately 100 feet above the ground. After running out of down elevator it stalled, fell off on a wing, and crashed, killing both occupants. Aft loading outside the approved center of gravity limits caused this accident even though the maximum allowable gross weight was not exceeded.

Forward CG Loading

One popular four-place low wing airplane built in 1981 and often used as a trainer is an example of forward CG loading—the opposite of aft CG loading. When this airplane is loaded with full fuel tanks and only two 170 pound persons aboard in the front seats, the CG is in front of the forward limit of the loading envelope! To bring the CG within the forward limit it is necessary to offload eight gallons of fuel.

With forward loading, nose-up trim is required in most airplanes to maintain level cruising flight, and this causes the airplane to cruise slower than it normally would with four persons aboard. To maintain altitude requires nose-up trim which produces a down load on the tail. Additionally, a higher angle of attack of the wing is required. This of course results in more drag.

A pilot who attempts to fly an airplane with forward loading might find it difficult to flare for the landing, especially if the approach was being made power-off, a little slow, or into a strong headwind.

Unintentional Spin In A DC-3

Former Air Force and FAA pilot Ronald R. Templin of Ft. Myers, Florida, describes an experience that he and another pilot had on a delivery flight of a Douglas C-47A (DC-3) enroute to Saigon. The airplane had been overhauled and was being delivered to the Vietnamese government. He writes:

"About halfway between Japan and Formosa, we flew—apparently too close—to the clear tail of an East China Sea typhoon. Cruising nonchalantly along on a sometimes erratic autopilot, there was a ripple of turbulence, then suddenly we were flipped into an inverted position. Below us, all that could be seen was the calm sea. With four busy hands in the cockpit, we managed to get the autopilot disengaged. In the interim, the long range fuel tank supply was interrupted by fuel starvation and both engines quit but kept windmilling. The aircraft stalled while we were rolling to the right, and as the nose fell, it dropped us into a spin. The airplane made a little over one full turn before we got it out of the spin. The whole experience lasted less than three minutes before we got back on course and climbed to our assigned altitude. We spent three days in Hong Kong recovering because the other pilot was pretty sick from a couple of good raps that he received on the head during that short period of panic we endured. We went over the aircraft with a fine-tooth comb, but did not find even one popped rivet. It is still a bad memory, but from then on I was a believer in Douglas Aircraft."

Opinions of an Experienced Pilot

One of the people who I contacted concerning this subject of stalls and spins was Floyd Wheeler of Louisville, Kentucky. You may recall from Chapter 5 Floyd's story of a student who was having trouble with spin recoveries in a Stearman. Floyd is a very experienced pilot in military (USAF) aircraft, and he has also spent many years flying all kinds of general aviation aircraft. Pilots who have flown with Floyd consider him the best. In fact, after the late Frank Wignall of Jackson, Mississippi, GADO gave him an Airline Transport Pilot (ATR/ATP) flight test in 1951, Frank told me it was the best check ride he had ever ridden through. His aviation experience includes: Instructor in PT-17, PT-19, BT-13 (Fig. 8-5), T-6, T-28; World War II pilot in P-40, P-51, B-17, B-29; Logair and charter pilot for AAXICO (Saturn) on the C-46 and DC-6; Corporate pilot on Cessna 310 and Beech Queenair 80. I would like to share with you some of his thoughts expressed in a recent letter. He writes:

"Most everything that I can come up with about stalls and spins is pretty much commonplace and very likely has been covered many times before, so don't hesitate to put this stuff I'm writing in File 13.

"I was over at Memphis about 15 years ago when a loaded Twin Beech lost an engine on takeoff. The pilot got himself into serious trouble and spun in on the final turn, killing all aboard.

"I am writing about stalls and spins during a single-engine go-around because I *think* this is what possibly happened to the Beech at Memphis. He made it to the final turn and thought there may have been other factors, had he been turning into the good engine it might have made the difference.

"Here are some of my opinions concerning stall and spin probabilities during a single-engine go-around or when losing an engine after takeoff in a twin-engine aircraft:

Fig. 8-5. 1942 photo of Floyd Wheeler beside a BT-13.

"All other factors being equal (such as obstructions, terrain, available runway for landing, etc.), it is far safer to turn toward the good engine rather than toward the inoperative one in a single-engine go-around. If enough rudder pressure could or would be maintained to offset the unequal thrust when turning into a dead engine it might not matter, but that is seldom the case. Instead, a critical loss of airspeed often occurs due to *slow* or incorrect performance of single-engine procedures. Then, as rudder effectiveness diminishes at lower airspeeds, slipping or skidding increases in magnitude, causing further losses in airspeed and a continued worsening condition.

"In a skidding turn, a stall is more likely to occur. It is far more treacherous and will occur quicker and at a higher airspeed than a stall in a slipping turn. In a skidding turn, the relative wind striking the outside of the fuselage causes increased drag and a reduced airflow over the inside wing. If a stall occurs, the greater lift on the outside wing or reduced lift on the inside wing usually results (with little warning) in a rather violent roll in the direction of the turn to a partially inverted position, along with a sharp downward pitch of the nose and minimal chances of recovery at a low altitude.

"A stall in a slipping turn is less severe and also less likely to occur. In a slipping turn the relative wind is striking the inside (direction of the slip side) of the fuselage, and while this does cause increased drag, it also produces a fair amount of lift, resulting in a lower stalling speed. The flow of air is somewhat restricted over the outside wing while the inside or low wing is producing greater lift. As a stall is approached, the bank will tend to decrease or the aircraft will slowly roll away from the direction of the turn, giving more warning and a far better chance of recovery with much less loss of altitude than would be required to recover from a stall in a skidding turn.

"If a twin-engine pilot fails to regularly practice simulated engine-out procedures under safe conditions, in order to develop a quick and automatic response to an engine failure during critical conditions, he would be safer with a single-engine aircraft. Sometimes that extra engine just gets pilots into a position that is far worse than being in the single-engine airplane they started out flying.

"Personally, I've tried various engine-out procedures and prefer this one over any of the other procedures, *when one engine fails on takeoff or during climb out.*

"If sufficient runway remains, close throttles and land. Otherwise:

"1. Maintain single-engine best rate-of-climb speed (straight ahead).
"2. Mixture, prop and throttles forward.
"3. Gear up.
"4. If flaps are down, milk them up.
"5. Maintain perfectly straight course with required rudder and iden-
    tify *good* engine.
    (Strong left rudder required—left engine good).
    (Strong right rudder required—right engine good).

"6. Bank into *good* engine 5-10 degrees.

"7. Maintain airspeed.

"8. Maximum power on *good* engine.
   (Same side as strong rudder pressure).

"9. Throttle back on inoperative engine and feather it.
   (Steps 5 and 8 should eliminate the chance of feathering the wrong engine).

"Other factors being equal, make pattern turns toward the *good* engine. If the airport is tower controlled, advise the tower of your intentions. Since the aircraft will be trying to turn toward the dead engine, simply remember to turn the hard way—*away from* the direction the aircraft wants to go.

"While turning toward the good engine in the pattern, if there is any loss of airspeed and the rudder becomes insufficient to counteract the unequal thrust of the good engine, a less dangerous slipping turn will result and a safe completion of the final turn will be more likely."

In 1951, an instructor at Columbus AFB, Mississippi, told me that he had ridden through a complete sequence of acrobatic maneuvers as Floyd Wheeler performed them under the hood in a T-6 on partial panel. I had also heard that while Floyd was on the Standardization Team at Foster Field in Victoria, Texas, he was performing outside loops in a P-40 (Fig. 8-6).

Since the P-40s were not equipped with an inverted fuel system, I was anxious to hear how he was able to perform outside loops in them. So I asked him to write me and explain how he performed acrobatics under the hood, as well as how the P-40 feat was accomplished. here is his reply which should be of interest to aerobatic pilots and instrument pilots:

"I've done outside loops in P-40s equipped with both Allison and Packard engines, neither of which had an inverted fuel system. Immediately upon assuming a negative wing load the engine of either one would cut out. I could never have completed an outside loop without power throughout the maneuver. I found that after roughly 15 seconds of inverted flight, the engine would surge back with full power, lasting long enough to complete the last half of an outside loop. (At the time, I asked several mechanics the reason for this, but never got a definite answer.)

"When I first attempted this maneuver, I practiced the last half first to find out just what airspeed was needed at the bottom in order to reach the top. I rolled the P-40 inverted at 200 mph, waited for the power to surge back in, then pushed the nose up and stalled out when nearly vertical. After several attempts, I found that 300 mph *plus* the surge in power was needed to complete the last half of the loop.

"Then I worked on the first half. Starting in straight and level flight, the nose was raised to about a 40 degree pitch attitude and slowed to just above a stall. Then the stick was briskly pushed forward as rapidly as possible without getting a negative stall. As the airspeed increased in the second

Fig. 8-6. P-40 of the type in which Floyd Wheeler performed outside loops.

quarter of the loop, by relaxing some of the forward elevator pressure to prevent a negative wing loading stall, I found the inverted position could be reached with as little as 200 mph. Since the speed needed for completing the last half of the maneuver was 300 mph, it was necessary to loosen up the loop (relax some forward pressure) as inverted flight was approached. The P-40 would soon attain 300 mph in the bottom of the loop, and the power surged back in about the same time. This provided enough airspeed and power to complete the loop.

"My experimenting with this and other outside maneuvers (outside snap rolls and inverted spins) wasn't all foolishness. I had ideas of going overseas in fighters, and after a goodly number of practice dogfights at Foster Field, I found that it was quite difficult to shake a good pilot off my tail with any conventional maneuver. But I could lose them instantly when I started doing outside loops. It wasn't necessary to do a complete loop, just pull into a tight nose-high turn and when the airspeed dropped enough, push forward briskly. There was no way they could see you and no way they could follow you.

"As you suggested, the acrobatics in a T-6 under the hood were done on partial panel; they were largely mechanical and a matter of timing. That sequence of maneuvers included spins, loops, snap rolls, slow rolls, barrel rolls, half rolls and reverse, inverted flight, and Immelmanns. For example, in a slow roll the airspeed and attitude was set up by partial panel, then the roll was strictly mechanical until the very end when the needle was used to level the wings. The same is true with a snap roll—mechanical and timing. In a barrel roll, you could tell where you were as you went along by watching the needle. In a right barrel roll, you start with a 20 or 30 degree left bank; the needle will swing left, then full right, then full left during recovery and back to center at completion.

"In an Immelmann, you know how tight to pull up by seat pressure. You also know what the airspeed should read when it's time to roll out.

"Inverted flight is like doing a back course ILS. It is necessary to roll in the direction the needle is deflected to center it, and the airspeed indicator is used for pitch control.

"I hope this is not too confusing and adequately answers your questions."

I remember one rather odd incident that happened to Floyd many years ago. An Aeronca C-3 received some damage to the elevators as it was being moved around in a hangar, and for this reason two mechanics removed the elevators to repair the damage. After the work was completed, Floyd was asked to test-hop the airplane. During the preflight the elevators appeared to move freely, so the C-3 was started and taxied to takeoff position. On takeoff, when he tried to raise the tail, it would not come up. Floyd closed the throttle and the airplane rolled to a stop. He looked back at the tail assembly and discovered that moving the stick forward raised the elevators and when the stick was back the elevators were down. Floyd taxied back to the hangar and told the mechanics to reverse the control cables at the elevator horn attachment points.

As Floyd points out, when you don't have an inverted fuel system it is necessary to improvise with extra airspeed, etc., in order to perform certain acrobatic maneuvers.

I will mention briefly how it is possible to modify the customary control usage for performing two acrobatic maneuvers in an airplane without an inverted fuel system. First, consider the Immelmann turn, which consists of the first half of an inside loop and the last half of a slow roll. The aircraft reverses direction 180 degrees. Performing this maneuver in a single-engine airplane in the usual manner without an inverted fuel system, we find the engine quits inverted during the roll-out. By making it a continuous maneuver without pausing inverted, it is possible to keep the engine running throughout the Immelmann and recover in a climbing attitude.

Here is the way it can be done: Using the same entry speed, the first part of the maneuver is identical to performing any other Immelmann. However, the last half of a slow roll type recovery is not used. Instead, for a roll-out to the right, when the nose of the airplane reaches a point between 50-60 degrees above the horizon, start applying right rudder which causes the nose to arc downward; simultaneously apply left aileron and a few ounces of back stick pressure. This slight back pressure produces enough centrifugal force (artificial gravity) to keep fuel flowing to the engine. As the nose arcs downward, the wings are rolled level with the ailerons. When the wings are within 45 degrees of being level, a slight amount of forward pressure is used with some left rudder. The airplane recovers in a climbing attitude with the engine running. I know it works, because I have done it a number of times in six different single-engine airplanes—the smallest was an Aeronca 7AC and the largest a T-28.

The other modified maneuver is the snap roll. Normally, the snap roll is actually a spin in a horizontal direction in which the airplane is stalled at a speed considerably above its normal stalling speed. It is usually entered from level flight with the abrupt application of full up elevator and full rudder in the desired direction of roll. The use of the ailerons is optional. I have always felt that doing a snap roll in this manner produced a sloppy-looking maneuver with the nose wallowing around the sky. The airplane is completely stalled on entry, and consequently, the controls are rather sloppy during the recovery.

The way I prefer to perform a snap roll is easier on the airplane and easy to accomplish. Depending on the airplane, this type snap roll can be entered at cruising speed or below cruising speed. With the nose in level flight or in a shallow climb, rapidly bring the stick back one or two inches (depending on the amount needed for a particular airplane). Immediately apply full left rudder and full left aileron for a snap to the left. The airplane should pivot on a point on the horizon while boring a tight small hole in the sky. Hold the controls in this position for approximately three-quarters of a turn, then apply recovery controls by relaxing the back pressure and using opposite rudder and aileron (from the direction of the roll). By refraining from stalling the airplane on the entry, it was not subjected to a heavy buildup of G forces. The recovery can also be made in a climbing attitude well above stall speed. Every airplane in which I have tried using the controls in this manner snap rolled very nicely.

RECOVERIES FROM SPIN ENTRIES

Getting back to stalls and spins in multi-engine airplanes, here are two incidents I recall witnessing:

When I was enrolled in a Turbo 680 T-Aero Commander familiarization course, it was customary for the instructor of this course to fly with two pilot trainees. Another pilot trainee and I would alternate flying from the left seat while the instructor occupied the right front seat.

The other pilot trainee was an experienced pilot of many hours of flying, although I observed he was often behind the airplane in many of the maneuvers he attempted. One day he was flying the airplane and I was riding in the seat directly behind him while the instructor had throttled back the left engine for a V_{mc} demonstration. As the airspeed decayed, I noticed the pilot trainee had the control wheel turned nearly fully to the right and the airplane's nose was starting to move slowly to the left. Since there wasn't anything I could do to stop what I saw developing, I started praying: "Father forgive him, he doesn't know what he's doing." A moment later, with the right engine still at full power, the Turbo Commander paid off with a wingover to the left. The instructor took over, brought both throttles to idle and popped the yoke forward. The rotation stopped after we had turned 90 degrees. At this point, the nose of the airplane was about 60 degrees below the horizon and the wings were flying again. The instructor returned to level flight. This was a case where an experienced pilot didn't recognize the loss of directional control, nor did he take prompt aggressive action to recover from a single-engine stall. I was pleased that the instructor didn't wait to see if the pilot trainee could recover from a spin.

During World War II, I was stationed at Bryan Field in Texas, at the Instrument Instructor school for the Air Force. We were using nine Twin-Beech AT-7s (which were similar to the C-45 or Beech 18) for Link trainer operator familiarization flights. These airplanes were equipped with desks and radio navigation instrument panels in the cabin area. The Link operator trainees would take turns in the copilot seat flying the airplane. One day a pilot and seven Link trainees were on a round robin flight in one of the AT-7s in instrument weather. Somewhere over Louisiana the right engine went out, and while going through engine-out procedures, the pilot failed to switch the engine-driven vacuum pump selector to the good engine. The vacuum pump selector valve was located on the floor beneath the copilot seat. Since the vacuum-operated flight instruments were not receiving vacuum they soon became inoperative or tumbled, causing the pilot to lose control of the airplane. The AT-7 then entered a diving spiral or spin and crashed, killing all aboard except one trainee who was in a rear seat near the cabin door. When he saw the altimeter indicating 1500 feet he pulled the door release and bailed out. It so happened that the pilot had less than 200 hours multi-engine time; he had been in combat as a single-engine fighter pilot and was rotated to duty back in the states.

As a result of this accident, the late Col. Joseph B. Duckworth, Commanding Officer at Bryan at that time and father of the "Attitude System" of instrument flying in the Air Force, asked me to study the problem and write a pilot procedures manual for the type flying we were doing with the Twin Beech. He suggested that I put 900 pounds of sandbags in the cabin area and record the stall speeds of the airplane in various configurations. I was also instructed to fly with the twelve pilots who were assigned to the AT-7s and work with them on extensive simulated engine-out procedures and stall practice both under the hood and out from under the hood.

On one occasion, I had a young lieutenant flying the airplane from the left seat—out from under the hood. We had the usual 900 pounds of sandbags tied down in the cabin floor. While attempting a stall in a steep left turn (45-50 degree bank) with lower than cruising power on both engines, he felt the horizontal tail assembly shaking. Since tail shaking occurs in the AT-7 well ahead of the stall, this pilot wanted to go beyond this phase with a minimum amount of shaking. He brought the control wheel back rather abruptly and the AT-7 rolled out of the bottom of the steep turn into a power spin to the left. I closed both throttles, popped the yoke forward and applied full right rudder. The rotation stopped immediately after we had spun a half-turn. Afterwards, I was glad this happened because both of us learned something from it. We also had more confidence in the Twin Beech since it responded nicely with the added weight in the cabin area. Even though this was a training flight, because of the aft loading I did not wait to see how the young pilot would respond with recovery controls.

MORE ABOUT FLAT SPINS

A book published in 1981 titled *Anatomy of A Spin* was written by John Lowery. If you are presently flying any type of modern general aviation airplane and haven't read this book, I suggest that you do so. (It is published by Airguide Publications, Inc. of Long Beach, California.) The following is an excerpt from the book concerning flat spins:

"Center of gravity is the first offender. An aft center of gravity causes an aircraft to spin with a flatter attitude. Even in the normal upright spin, an aft CG delays spin recovery. The reason, as discussed earlier, is that the aircraft goes deeper into a stall or stalls more completely. With an aft CG at high angles of attack, the available down elevator force is diminished, yet more down elevator is needed to break a stall or spin.

"Let's say that with an aft CG, as with a load of passengers on a flight demonstration, we inadvertently encounter an abrupt stall. To counter a left rolloff tendency, we unconsciously apply aileron control against the roll. The adverse yaw from the downward deflection of the left aileron, combined with high engine power and P-factor, provide the ingredients of a flat spin to the left.

"The second great offender in either twins or singles is a high power setting. In many light twins, with one engine at idle simulating an engine failure and the other at full power, a loss of control requires that both engines be brought promptly to idle. As the 'departure' progresses toward a steady state spin, a high power setting brings the nose up and leads to a flat spin."

I concur fully with author John Lowery's opinion expressed in the introduction of his book. It states in part:

" . . . I feel strongly that spin training should be required before licensing a private, commercial or instructor pilot. With 28 to 48 percent of our General Aviation fatalities attributed to stall/spin problems, it is obvious that more training is required. Many flight instructors show a great weakness in both knowledge and proficiency in this area. This weakness is thus passed along to students.

"The basic aerobatics course available from several fixed base operators emphasizes spin training which makes it an invaluable investment to pilots of all categories. The training this type of course provides is like life insurance and every serious pilot and flight instructor should attend one."

Chapter 9

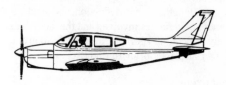

Significant Factors
in Airplane Spinning

There has always been a spin problem, starting with the first airplanes that were built. In the early days there were good spinners and bad spinners. The same is true today, as we have airplanes with good spin recovery characteristics while others have rather weird characteristics in a spin. In 1935, full-scale spin tests were conducted on several airplanes by the National Advisory Council on Aeronautics (NACA), and the results from these tests led to the so-called NACA-recommended spin-recovery technique: Briskly move the rudder full against the spin; after a lapse of appreciable time (approximately one-half turn), briskly move the elevator to approximately full down, and hold these controls until recovery is complete. The airplanes of that day probably were in the zero loading condition, and today this recovery technique would apply only for airplanes that have similar loadings.

NASA STALL/SPIN RESEARCH

The National Aeronautics and Space Administration (NASA) Langley Research Center has initiated a broad general aviation stall/spin research program, which consists of spin/tunnel tests, radio-control powered-model tests, and full-scale flight tests. This program was designed to study and evaluate the spin characteristics of configurations typical of light airplanes flying today. As of June 1, 1979, it was necessary to deploy the spin recovery parachute 16 times while performing 529 spins in one of the low-wing research airplanes used in the tests.

Most of the material in this chapter has been excerpted from NASA technical notes and papers written by James S. Bowman, Jr., and his associates, Sanger M. Burk, Jr. and William L. White of the NASA Langley

Research Center, Hampton, Virginia. Highlights of this subject matter were selected at random from three NASA technical notes and papers, and the illustrations that accompany the material are reproductions of those appearing in the technical papers. The following are excerpted quotes:

"Factors which influence the spin characteristics of light airplanes are quite numerous, and a list of only a few of these factors would include mass distribution, relative density, wing position, center-of-gravity position, tail configuration, tail length, aileron deflection, and fuselage shape. Because of the large number of variables involved and the lack of research in this area, it is extremely difficult and perhaps impossible to predict the spin characteristics of light airplanes prior to flight tests.

MASS DISTRIBUTION

"The airplane mass distribution has an almost overriding effect on the spin and spin recovery. The way in which the mass of an airplane is distributed between the wing and fuselage is the most important single factor in spinning because it determines the way in which the airplane, while spinning, responds to control movements, especially to elevators and ailerons. An airplane rotating in a spin can be considered to be a large gyroscope. Since there are mass angular rotations, about all three axes, inertia moments are produced about all three axes. In addition, aerodynamic forces and moments are acting on the airplane because of its motion through the air. The developed spin involves a balance between the aerodynamic forces and moments and the inertia forces and moments acting on an airplane.

"The mass distribution of all airplanes can be grouped into three general loading categories, as indicated in Fig. 9-1. Shown at the top is the case of an airplane that has most of its weight distributed along the fuselage. This situation is referred to as *fuselage-heavy loading*. Features such as fuselage-located engines, fuel, luggage, and cargo contribute to such a loading.

"Almost all light general aviation airplanes, even light twins, fall into the *zero loading* category as shown in Fig. 9-1. However, there are some exceptions, especially when heavy tip tanks are installed on the wings.

"When the weight of the airplane is distributed mainly along the wing, this is called *wing-heavy loading*, and features such as wing-mounted engines and tip tanks contribute to such a loading.

"The loading of an airplane dictates the control movements required for recovery. Deflection of the rudder to oppose the spinning rotation directly is always recommended, but in many cases it is not adequate to provide recovery. For the *wing-heavy* loadings, down-elevator is the primary recovery control. For *fuselage-heavy* loading, the aileron is the primary recovery control; the aileron should be deflected with the spin—for example, stick right for a right spin. For the *zero loading*, the rudder is always an important control for spin recovery. Therefore, any airplane in this loading

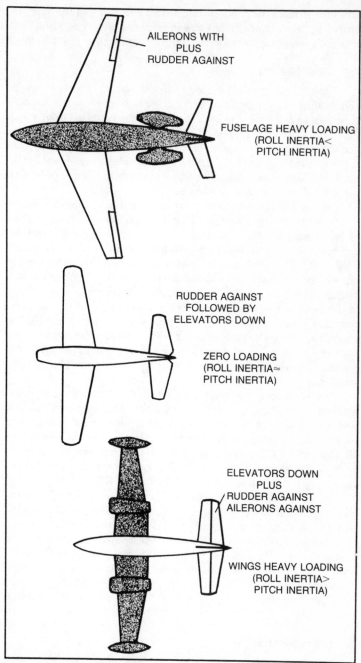

AILERONS WITH
PLUS
RUDDER AGAINST

FUSELAGE HEAVY LOADING
(ROLL INERTIA<
PITCH INERTIA)

RUDDER AGAINST
FOLLOWED BY
ELEVATORS DOWN

ZERO LOADING
(ROLL INERTIA≈
PITCH INERTIA)

ELEVATORS DOWN
PLUS
RUDDER AGAINST
AILERONS AGAINST

WINGS HEAVY LOADING
(ROLL INERTIA>
PITCH INERTIA)

Fig. 9-1. Primary recovery controls as determined by mass distribution.

condition should have a rudder designed for effective spin recovery. Determining the elevator effectiveness in the zero loading range is difficult, especially where the 'zero loading' tends toward fuselage-heavy loadings.

TAIL CONFIGURATION

"The tail configuration is a very important factor in the zero or near-zero loading range, where the rudder is the primary recovery control. A relatively large moment is needed to recover an airplane from a spin, especially a flat spin. Therefore, it is important that the airplane control surfaces, particularly the rudder, be effective at spin attitudes. The problem is that during a spin, much of the rudder usually is in the stalled wake of the horizontal tail and sometimes the wing, over which the dynamic pressure is low or abnormal airflow conditions exist.

"Figure 9-2 shows the factors which are important in the tail configuration for spin recovery. This figure illustrates the dead-air region over much of the vertical tail, which is caused by the stalled wake of the horizontal tail and which seriously decreases the effectiveness of the rudder. In order to have good rudder effectiveness, a substantial part of the rudder must be outside the horizontal tail wake. Another important, but less obvious, consideration is that the fixed area beneath the horizontal tail be sufficient to dampen the spinning motion, since it has been found that this area contributes much of the dampening of the spin rotation (Fig. 9-3). The rudder must have a substantial amount of area outside the horizontal tail wake in order to be effective, and also the fuselage must have a substantial amount of area under the tail in order to provide damping of the spinning rotation.

"Based on rotary balance test results obtained from the spin tunnel within the past few years, the relative position of the horizontal tail (HT) with respect to the vertical tail (VT) is significant and can cause the airplane to have more dampening, or more propelling in a spin, depending on how the HT shields the VT. In general, the most dampening moment you can get in a spinning attitude is with the HT off. The T-tail configuration is about as good for spin dampening as HT off configurations (Fig. 9-4).

"Some airplanes were designed with small tail assemblies and the tail was just barely satisfactory from a spin-recovery standpoint. Then over the years, the airplane's gross weight is increased because of design changes or changes in operating practices. If some of this weight increase were in the form of wingtip tanks or heavier engines on the wing, the airplane's characteristics would be even more unsatisfactory. Such changes in spin-recovery characteristics from satisfactory to unsatisfactory have actually happened in the past as a result of a weight increase over that of the basic airplane design.

Rudder Effectiveness

"In general, two types of rudders are used on general aviation airplanes today—full-length rudders and partial-length rudders. Full-length

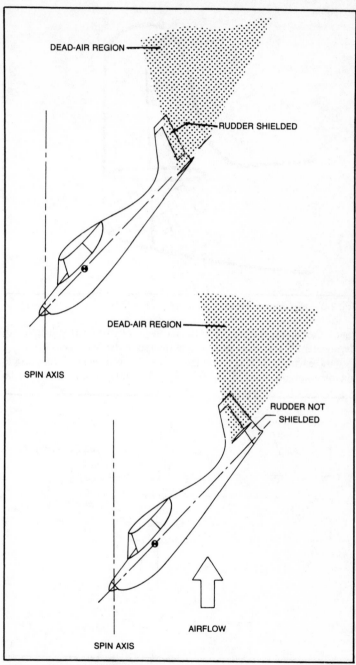

Fig. 9-2. Effect of horizontal tail shielding on rudder.

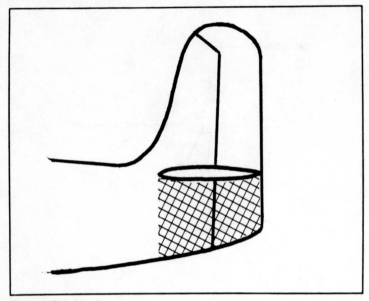

Fig. 9-3. Tail-damping power.

rudders extend to the bottom of the fuselage, whereas partial-length rudders generally terminate at or above the top of the fuselage.

"The 'tail dampening power factor' has been proven invalid and should not be used. As pointed out in NASA TP 1009, it was shown to be

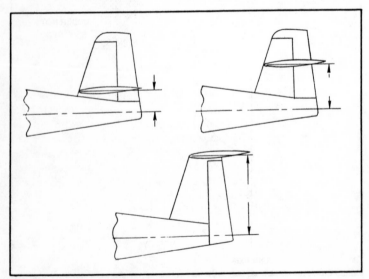

Fig. 9-4. Horizontal tail configurations.

124

inadequate to predict spin recovery. Based on rotary balance testing for several general aviation designs, the fuselage and wing contribution to the characteristics of the spin can often times overpower the tail contribution.

"In general, the optimum horizontal-tail position to provide the maximum unshielded rudder area is different for partial-length and full-length rudders.

"For full-length rudders, the part of the rudder below the horizontal tail provides most of the unshielded area. Therefore, high and forward positions of the horizontal tail are usually the most effective configurations for spin recovery for designs employing full-length rudders.

"For partial-length rudders, all the rudder above the horizontal tail and the top part of the rudder provides most of the unshielded area. Therefore, low and rearward positions of the horizontal tail are most effective to provide the needed unshielded rudder area.

"One particular point that should be recognized with regard to tail design is that with a low horizontal tail and a sweptback vertical tail, it is possible that almost the entire vertical tail, including the rudder, might be in the stalled wake of the horizontal tail.

"*Tail length* can have an appreciable effect on the spin and recovery characteristics of an airplane. Tail length is generally expressed nondimensionally as the ratio of the distance between the center of gravity and the rudder hinge line to the wingspan. The general effect of increasing the tail length is to cause the airplane to spin at a lower angle-of-attack and at a higher rate of rotation, whereas the shorter tail will cause the airplane to spin flatter and at a slower rate of rotation.

Elevator Effectiveness

"For airplanes with partial-length rudders and often the full-length rudders with a low horizontal tail, the rudder is usually mostly shielded by the horizontal tail and is consequently ineffective for spin recovery. Therefore, the elevator is relied on for most of the spin recovery. Even so, an almost universal control technique suggested for recovery is rudder reversal followed by deflection of the elevators to neutral or down. Because most light general aviation airplanes are required in the spin demonstrations to rotate only one turn before recovery attempt, this technique is usually successful, provided that recovery is attempted before one turn is completed. However, in many cases it would be disastrous for the airplane to inadvertently wind up more than one turn because this technique may not recover the airplane if the spin has developed to two or more turns. Normally, a light airplane does not attain an equilibrium or balanced spin condition until after approximately two to five turns.

Wing Position

"The position of the wing (high or low) is believed to have some influence on the spin and recovery characteristics of airplanes. The history of stall/spin problems associated with high or low-wing airplanes indicates

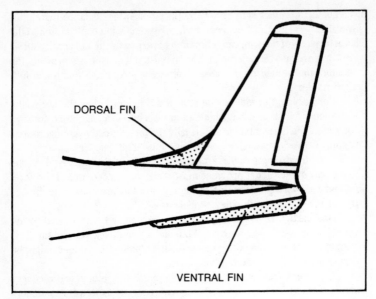

DORSAL FIN

VENTRAL FIN

Fig. 9-5. Dorsal fin and ventral fin.

that a high-wing airplane is expected to have better spin and recovery characteristics than a low-wing airplane, all other factors being equal. The reason for the apparent improvement in spin characteristics of the high-wing airplane is believed to be related to the higher dihedral effect caused by the high-wing position and to improvement in the wake characteristics of the wing in the vicinity of the tail. The wake from a high wing is believed to pass above the tail so that the tail surfaces are not appreciably affected and the rudder and elevators are more effective in the spin recovery.

Spin Tunnel Tests

"An investigation has been conducted in the Langley spin tunnel on a model of a research airplane which represents a typical low-wing, single-engine, light general aviation airplane. Some of the results of these tests are:
"Modifications to the fuselage such as rounding the fuselage bottom of adding ventral fins at the tail were found to be effective in eliminating a flat spin condition (Fig. 9-5). For the configuration tested, a very sensitive airflow phenomenon existed at the trailing edge of the wing at its juncture with the fuselage with the result that wing fillets, corner modifications, and ventral fins located in that area resulted in dramatic changes in the spin for some tail configurations (Figs. 9-6 through 9-8).
"The results also showed that the aileron setting during the spin had a marked effect on spin characteristics such that aileron deflection in the adverse direction could have a strong adverse effect on recovery.

126

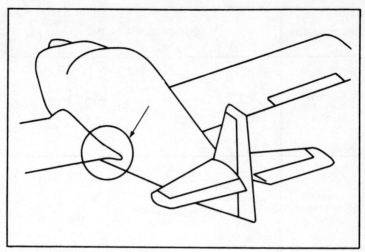

Fig. 9-6. Wing trailing edge fillet.

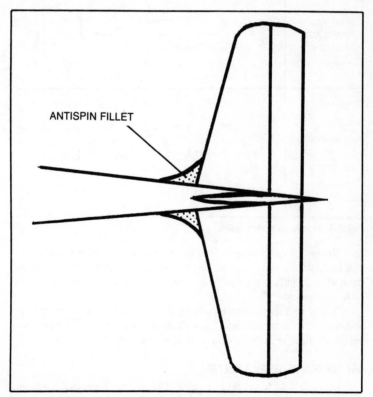

ANTISPIN FILLET

Fig. 9-7. Typical tail design showing antispin fillets.

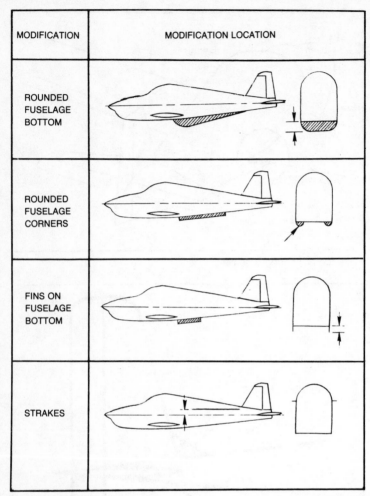

MODIFICATION	MODIFICATION LOCATION
ROUNDED FUSELAGE BOTTOM	
ROUNDED FUSELAGE CORNERS	
FINS ON FUSELAGE BOTTOM	
STRAKES	

Fig. 9-8. Modification examples.

"Investigation of the effects of tail configurations on the spin-recovery characteristics showed that the T-tail and the tail configurations that had the horizontal tail mounted halfway up on the vertical tail produced the best spin recoveries.

"Many geometric features other than tail configuration affect the spin and recovery characteristics. Some of these features are fuselage cross-sectional shape (Fig. 9-8), ventral fins, strakes, and wing fillets."

ARE WE OUR BROTHER'S KEEPER?

Omer Broaddus, a flight instructor of Lexington, Kentucky, has been a friend of mine for over 40 years. He told me of a spin-in accident that he witnessed many years ago at a small airport. This accident involved a student pilot with approximately five hours solo flight time who had purchased a new Aeronca K. The day this accident occurred, a salesman from the Aeronca factory landing at this airport with an Aeronca K demonstrator. To attract attention, the salesman entered the traffic pattern and flew a close-in downwind leg along the airport boundary. Upon entering the turn to base leg, the salesman started a steep slipping turn (turning 180 degrees) into the airport. He continued slipping on final until leveling the wings just before touchdown.

The student pilot observed the salesman's spectacular performance and decided to try the same type landing approach in his airplane. Without discussion slipping turns with anyone, the student proceeded to take off with his brother as a passenger. When they returned to the airport the student also flew a close-in downwind leg in order to duplicate the salesman's performance. During the turn to base leg, a steep banked nose-high power-off turn was started. Shortly thereafter, the airplane entered a spin; both occupants were killed instantly when the airplane struck the ground in a near-vertical attitude.

I am afraid that at times I have also been guilty of not always setting a good example for inexperienced pilots. On many occasions while demonstrating airplanes to prospective customers, I have pulled them off the ground after a short takeoff roll, followed by a maximum performance climb to the airport boundary. In many cases, the airplanes didn't really have good climb performance, yet I tried to make it appear as though they would go up like an aluminum elevator. Of course, the steep climbs were always done straight ahead with no turning flight, yet what impressions or ideas did I give the young pilots who were watching me? Climb too steeply after takeoff looks easy and it is one way to show off with an airplane. Yes, I believe that we are or should be our brother's keeper. Imagine how I would have felt if later one of my students had spun in trying to duplicate my takeoffs.

Omer Broaddus supplied most of the old airplane pictures that appear in this book. He is one of those wise fellows who always carries a camera wherever he goes. I remember attending a breakfast flight at a small town airport many years ago, and Omer wanted to take aerial photos of the parked airplanes. He asked my to fly him in a J-3 Cub on some low passes around the airport for the close-up shots. The right side of the J-3 cabin can be opened in flight, which makes a good arrangement for shooting pictures. I was flying the J-3 from the front seat with Omer and his camera in the rear seat, although we did not remove the rear seat dual controls prior to this flight. After making several low-level passes, Omer had finished taking pictures so I entered a climbing turn. When I tried to lower the nose, the control stick would not move forward because his large Graflex camera was wedged between the rear stick and the back of the front seat. I yelled to him that I would bring the stick back more if he would kindly remove the camera at the same time.

This incident has been related to show how dual controls might interfere with taking pictures from an airplane. On photo fights, it is usually necessary for the pilot to fly fairly slow in a banked attitude so the airplane's wings, struts, etc., won't appear in the photographs. Also, cameramen have been known to step on a rudder pedal while trying to position themselves to shoot a picture. Removing the dual controls before flying a photographic mission could prevent a stall or spin accident.

MICROJET SPIN

Airshow pilot Bob Bishop is a native of Lubbock, Texas, although he left there at an early age because his father was an airline pilot based in Phoenix, Arizona. Presently, he is living at Edmond, Oklahoma (in the Oklahoma City area). Bob is primarily in the airshow business; he travels around the country giving performances in his Acrojet and 300-hp Bellanca Viking, and is also involved in making television commercials.

During lunch one day, Bob told me of his experiences in spin testing a Microjet 90, which is a modified Bede BD-5J jet (Fig. 9-9). The spin tests were being conducted in 1976, starting at 13,000 feet AGL. He had been doing spins, both left and right, a half turn buildup at a time, and he had no problems in working up to two and one-half turns. Everything went well until he attempted a three-turn spin to the left, applying recovery control inputs after two and one-half turns. The small jet immediately entered a flat spin, rotating at a rate of one turn per second, with the nose approximately 20 degrees below the horizon.

Bob held the recovery control inputs for 10 turns but there was no change in the flat spin. He tried every recovery technique he could think of including lowering the flaps and gear plus adding full power. With the altimeter indicating 5000 feet AGL (it was set at zero before takeoff), he tried to release the canopy and this required the use of both hands. When he wasn't holding the stick it moved all the way aft and stayed there while he opened the canopy. The canopy on the Microjet is hinged at the rear and normally opens upward and to the rear, but since air was blowing underneath it, the canopy remained directly above the cockpit. After pushing the canopy back, Bob chose to jump to the outside of the spin to keep from striking the tail assembly. He felt very little centrifugal force while in the flat spin because the pilot's seat is very close to the center of gravity of this airplane. However, as he climbed out of the cockpit with one foot on the wing, the centrifugal force was so great that he fell on his back and landed on the top of the wing. He was soon free of the airplane by sliding off the wingtip. A chase plane pilot reported that the Microjet made 40 turns of a flat spin before Bob left it.

Bob delayed opening his parachute until he was well below the Microjet and then tried to track away from it. During the descent he could hear the airplane but never saw it until it hit flat in a pond. His parachute came down fairly close to the airplane and Bob broke an ankle as he landed.

Bob Bishop mentioned that a high spin-rate *fuselage-loaded* airplane like the Microjet in a spin is similar to a large gyroscope, and you are no longer dealing entirely with primary aerodynamics, but instead with a fundamental property of gyroscopic action called *precession*. Precession is the resultant action or deflection of a spinning wheel when a deflective force is applied to its rim. When a deflective force is applied to the rim of a rotating wheel, the resultant force is 90 degrees ahead in the direction of rotation and in the direction of the applied force. In other words, deflecting the ailerons or using down-elevator might produce a resultant action 90 degrees in the direction of rotation, etc. He also mentioned the ailerons are often the primary spin recovery control in fast spinning aircraft with fuselage-heavy loading (inertially).

At one time, Bob was an airshow demonstration pilot for Bellanca performing in the Bellanca Super Viking. The Viking shown in the cover photo going straight up is his new airplane in which he gives an impressive performance of vertical rolls, loops, hammerhead turns, slow rolls, double snaps on top of a loop, etc. He had also spun this airplane.

The spectacular aerial photo of Bob's Viking on the front cover of this book was made by Jim Larsen of Robertson Aircraft. The photo was made near Everett, Washington, by Jim Larsen from the rear seat of a T-6 with the rear canopy open, while the Viking and T-6 were in a formation loop.

OX-5 SPIN STORY

One of the first people I interviewed before starting on this book was Warren Smith, former FAA Chief of Flight Standards Training Program at the Aeronatical Center in Oklahoma City. For five years prior to his retirement in 1972, Warren served as Asst. Superintendent of the FAA Academy. He started with CAA in 1939 as a GADO Inspector in Kansas City, Missouri.

Warren told me that in 1936, while operating a flying service in the Pacific northwest, he observed one spin accident that he will never forget because it involved a close friend of his. At that time, two young men bought

Fig. 9-9. Bob Bishop's Bede BD-5J jet. (courtesy Bob Bishop)

an old OX-5 Swallow and re-covered it in his shop. Upon completion, they asked Warren if he would spin test if for them, as the Swallow had a reputation of being a bad spinner. In fact, CAA put on restrictions prohibiting intentional spins in the Waco 10, Swallow, American Eagle, and several other airplanes during the middle 1930s.

Warren was wearing a parachute while he spun the Swallow one turn to the left and one turn to the right. The nose started to come up during both spins and each time he applied full opposite rudder and full down elevator, recovering before it locked in a flat spin. He cautioned the owners that they should never attempt spins in the Swallow because of its characteristics.

Later the young men were talking to another who volunteered to fly the Swallow and do more spin testing in it for them. This 26-year-old pilot was a handsome, dashing young aviator. He had about 100 hours of flight time and held a Limited Commercial license. Warren had taught him to fly and said he was an excellent pilot. When the Swallow owners told this pilot that Warren had advised against spinning the airplane, he replied, "Smith thinks he is the only one who knows how to spin an airplane."

One Sunday morning Warren went to visit another friend who had a home in the country. While he was there he saw the Swallow climbing to altitude, and watched as the pilot kicked it off the right, making about six turns of a spin before recovering. Warren turned to his friend and said, "Well, that should be enough to satisfy him." But it wasn't. The pilot climbed back to about 5000 feet and entered a left spin. This time the airplane went into a flat spin and Warren counted 14 turns before it hit the ground. The pilot jumped out about 150-200 feet above the ground but didn't have time to open his chute.

When CAA was proposing to eliminate spin training from pilot certification requirements, Warren argued against it. Several years ago, his son Chuck was enrolled in a flight training course, and the instructor told Chuck that he did not teach spins. Before Chuck soloed, Warren rented a Cessna 150 and gave his son spin training.

More recently, a private pilot friend of his with no experience in spins purchased a Luscombe. Warren felt his friend would be safer in that airplane if he was familiar with spins, and for that reason, he volunteered his services for spin training in the Luscombe.

After Warren read my *Private Pilot* Magazine article "To Spin or Not To Spin—That's The Question," I asked him what he thought about the opinions I expressed about weaknesses in present stall recovery procedures. Here is his reply:

"I agree that the ailerons may be getting some pilots in trouble when their airplanes are fully stalled. I also believe in aggressive use of the elevators when recovering from stalls. The expression 'pop the stick' doesn't bother me in the least. Often it may be the only way to go.

"I still feel that every airplane pilot should experience spins as a part of his training."

Chapter 10

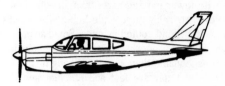

What Is the Solution?

Newspaper accounts of spin accidents all read about the same, and they include statements such as: (1) The pilot was known to be an expert pilot *or* the pilot was usually described as an expert pilot; (2) An eyewitness said the plane was circling before it dove straight down; (3) The pilot was seen circling the home of a friend.

NTSB reports of the same accidents often list as a probable cause one of the following:

☐ Pilot in command failed to maintain flying speed.

☐ Dual student failed to maintain flying speed, and the pilot in command inadequately supervised the flight.

☐ Pilot in command failed to follow approved procedures and directives (aircraft placarded against intentional spins).

☐ Aircraft entered flat spin during stall demonstration, unable to recover. The pilot in command executed improper operation of the flight controls.

☐ Improperly loaded aircraft.

I have listed several similar statements of my own. Could some of these be causative factors in the spin accidents?

☐ Pilot in command's stall recovery technique caused airplane to spin.

☐ Pilot in command did not receive proper training and was unaware of the extremes in behavior of a fully loaded airplane.

☐ Pilot in command did not realize airplane was on the threshold of a spin entry.

ARE SMALL AIRPLANES SAFE?

Not long ago a non-pilot friend asked me these two questions: "Are small planes as safe as airliners? Why are there so many crashes of small planes?"

I tried to answer him by saying, yes, small airplanes are as safe or safer than large jets or even automobiles. In the first place, airplane engines are superior to automobile engines; they are precision-built of the best material available, and with fewer moving parts than an automobile engine. Dual ignition improves their reliability, while being air-cooled eliminates the plumbing required for liquid-cooled engines. Small airplane engines, when operated and maintained properly, are as dependable as jet engines and a great deal more dependable than automobile engines.

The airplanes of today are fast and efficient and their takeoff and landing performance is good. In general, they are sturdily built and safe if operated in accordance with the operating limitations contained in the pilot's operating handbook or owners manual.

An airplane is as safe as the pilot who flys it, just as an automobile is only as safe as the person driving it. When a car traveling at 55 mph crosses the center median and hits another car head-on, we can't blame the accident on either of the cars involved.

Just how safe a pilot may be depends to a great extent on the training he received, and today good flight instructors seem to be a scarce item. Flight instruction is not a lucrative business, so many young pilots use it as a stepping-stone to better paying jobs in aviation. It is also a fact of life that all good pilots do not develop into good flight instructors. In training flight instructors over the years, I found that out of ten instructor applicants usually about three of them become excellent flight instructors; while some of the others had little imagination or dedication, and consequently never developed beyond safety pilots who protect the aircraft while students learn to fly by themselves.

After World War II and again after the military contract schools closed during the early 1960s, there was an abundance of good flight instructors throughout the country. Most of these pilots are no longer in the flight instruction business, which has resulted in a shortage of good, experienced flight instructors. Watered-down procedures and weak training of flight instructors hasn't made general aviation safer in recent years. It won't be easy to correct 32 years of retrogression or neglect in pilot and flight instructor training, but my opinion is that upgrading of flight instructors must first be accomplished before any real progress in flight training will be made.

I told my friend that I would not ride in an airplane with any present-day pilot unless I have access to a set of dual controls. I do not care to experience the helpless feeling of being with them as a spectator in an inadvertent stall or spin.

After our coffee cups were refilled my friend and I resumed our discussion about the safety of small airplanes. I continued with my views of the problem:

Today the people who are learning to fly and buying airplanes are the successful business and professional men. Their time, they know, is valuable; they have become accustomed to success and often treat their airplanes the same way they operate an automobile. That is, they kick the tires and climb aboard; if they can get it started, they fly it. Some of these pilots were oversold on instrument flying so they obtained an instrument rating, thinking that would qualify them to go anywhere anytime with their airplane.

Business and professional men often move up rapidly to larger, faster, and more sophisticated airplanes. After obtaining a private pilot certificate, they soon learn about their equipment limitations and how inclement weather keeps them on the ground many times. To overcome these shortcomings, they have more equipment installed in their airplanes or purchase larger airplanes equipped with DME, transponder, a flight director/autopilot system, etc. What some of these pilots don't realize is that they still can't operate like the airlines, yet they fly at night in instrument weather conditions and often attempt instrument approaches at airports they have never seen in daylight. Of course, airline pilots are thoroughly familiar with the routes they fly day-in and day-out, and they also have two of everything (i.e., pilots, radios, and always more than one engine). Every time a fast-moving cold front progresses eastward across the country, it usually takes with it some of these private pilots who get trapped attempting an approach when low ceilings, fog, snow, thunderstorms, or icing conditions are present. Just by reading the newspaper accounts of aircraft accidents in various states, you can often tell the progress a cold front has made across the country. I concluded my remarks by saying, "Yes, small airplanes are very safe—*but.*"

NO SPIN-IN PROBLEMS

When I think about airplanes of the late 1930s and 1940s, I remember what a joy they were to fly. Examples of several fine old airplanes are shown in Figs. 10-1 and 10-2. To my knowledge, none of them had any spin problems or a reputation for spinning in. On the other hand, I wouldn't care to spin any of them will a full load of passengers. The reason I believe we never heard of spin problems with these old airplanes is that the pilots flying them were familiar with spins. They also knew when they were nibbling at a stall or flirting with a spin, and they took prompt aggressive measures to fly out of those situations. Pilots flying the older airplanes were accustomed to performing spins, usually in airplanes smaller than those shown in the photos; thus, they could recognize what an airplane was going to do before anything unexpected took place.

Some of the older airplanes such as the Stinsons and Howards were built almost more sturdily than they needed to be, which meant they were

Fig. 10-1. Fine old airplanes of the late 1930s and early 40s. From top: Beech Staggerwing, Howard DGA, Waco cabin. (courtesy Omer Broaddus)

rather heavy. During the 1940s, just about every large airport had a hangar that housed one or two old heavy Brand-X hangar queens. For years, these no-purpose all-concrete monoplanes remained back in the corner of a hangar collecting dust, and the highest they ever got was when a still in the next hangar blew up.

Fig. 10-2. More fine old airplanes. From top: Stinson Reliant SR-9, Waco cabin, Monocoupe 90-F. (courtesy Omer Broaddus)

ANOTHER METHOD OF RECOVERY

Throughout this book, I have mentioned the pitfalls for pilots attempting to raise a low wing using a combination of back pressure and ailerons when the wings are stalled. I also believe a number of airplanes went into flat spins because *back pressure* and *opposite aileron* (from the direction of rotation) was used at the start of an inadvertent spin.

Most manuals and pilot's handbooks recommend that the ailerons be neutralized when intentional spins are practiced, which is another way of saying *don't use the ailerons and full up elevator travel at the same time in spins!*

In 1941, when I was instructing in Stearmans at a primary contract school, the commanding officer there had some different ideas about spin recoveries. It was his policy that the ailerons be neutralized while performing intentional spins. However, he insisted that we teach the use of full *opposite aileron* on spin recoveries. In other words, to recover from a spin to the right, we applied full left rudder, followed by full down elevator and full left aileron (the stick was in the left front corner). The commanding officer at that school was considered a very sharp and knowledgable pilot, yet this was the only military flight training school that I ever heard of which used this spin recovery technique. I do not remember whether or not this method of recovery stopped the spin rotation in a Stearman sooner than the conventional rudder/elevator method.

A good friend of mine claims that he has spun empty DC-3s several times, and found the use of ailerons against the spin on *recovery* also worked best in those airplanes.

We know the combination of crossed ailerons and back pressure is considered poison in many airplanes when applied during the incipient or developed stages of a spin. However, during a spin recovery in other airplanes (especially those with wing-heavy loading), ailerons and rudder applied against the direction of spin rotation plus down elevator may be the most effective recovery technique. The ailerons and rudder are not crossed when used in this manner—both are against the direction of rotation.

This matter of a proper use of the controls in spins and proper spin recovery techniques has many gray areas. It is not all black and white for airplanes since the correct procedures vary for different airplanes. Pilots must abide by the manufacturer's recommended procedures contained in the owner's manual or Pilot's Operating Handbook for each airplane they fly.

CLEARING THE AREA BEFORE SPINNING

Before attempting spin practice or any acrobatic maneuvers, the pilot must find a place that is off airways in uncontrolled airspace. Whatever clearing is deemed necessary to ascertain that the area below you is clear of other aircraft should be made before performing any maneuver (Fig. 10-3). For several decades, military flying schools have taught their pilots to perform at least one 180 degree clearing turn in each direction before entering such maneuvers as spins, Cuban eights, Immelmanns, etc., where considerable altitude changes are involved. There should be no delay in entering a maneuver upon completion of the clearing turns. This can be accomplished by performing the necessary conditions of flight (reducing airspeed, adding carburetor heat, etc.), while in the clearing turns. Remember, the sky is full of Lock Haven and Wichita flak. Don't get hit by any of it.

I remember an instructor in a primary contract flying school complaining about having a cadet who always forgot to look around or clear the area. The instructor told everyone that he intended to recommend the student for elimination from flight training for failure to look around. During graduation exercises, another instructor observed that this student was one of those in the graduation class. When the student's instructor was asked why this cadet was not eliminated, the instructor replied: "I decided to go ahead and put him through the program. Although I know he'll never be good enough to shoot down an enemy aircraft, I figure he's worth saving because he might run into one of them."

OWNER'S MANUALS

As stated previously, vigorous and immediate recovery action is required to avoid inadvent stalls and spins. This makes it appear that most aircraft accidents result from pilot error. Unfortunately, the owner's manuals and Pilot's Handbooks for certain airplanes described stall characteristics as "gentle" or "normal" (whatever *that* means). In real life, this is not always true if the airplanes are fully loaded.

When incomplete or misleading information is contained in the aircraft owner's manuals, the unsuspecting pilot may not be able to cope with an emergency as rapidly and accurately as he should. We rely on the airplane to behave in a "sensible" way—the way in which we have learned it should behave, the way which made us feel safe in it. Surely there would be fewer

Fig. 10-3. This Stearman was performing a chandelle when it collided with another Stearman recovering from a spin. Both landed safely, this one with one elevator.

fatal accidents if emphasis was placed on the potentially catastrophic and often irreversible situations which may develop, such as the flat spin—if a loss of control is allowed to occur in airplanes prohibited from intentional spinning.

It should be remembered that airplanes certificated under Normal Category rules have not been tested for more than a one-turn or three-second spin, and their performance characteristics beyond these limits are unknown. This is why they are placarded against intentional spins. The pilot of an airplane placarded against intentional spins should assume that the airplane may become uncontrollable in a spin.

If there is a lack of complete information in the manuals concerning faults and limitations of the aircraft, pilots operate them with a false sense of security and may find themselves acting as test pilots in their airplanes.

DISTRACTIONS

The FAA General Aviation Pilot Stall Awareness Training Study emphasized how stall/spin accidents are usually caused by a distraction of the pilot from his primary task of flying the aircraft. Here are several examples of what I consider pilot distractions:

☐ Passengers who carry a continuous line of chatter, or those who ask the pilot questions during takeoffs and landings.

☐ When six or more radio frequency changes and radio contacts are required to enter or depart an airport, the pilot may overlook a vital aircraft operational task.

☐ A pilot with an emergency should not be plagued by a controller asking too many questions (that will look good in his accident report).

SPIN TRAINING

Pilots without spin training flying spinnable airplanes, will always have hangups about spins because they fear the unknown. This is easy to overcome by flying with an experienced instructor in an airplane such as a Bellanca Decathlon or Citabria, or Cessna Aerobat, or a similar airplane that is certified for spins. There are a vast number of aircraft no longer in production which are certificated as either fully acrobatic or acrobatic with certain restrictions. This includes many ex-military airplanes and also antiques and classics. Older airplanes in particular should be carefully scrutinized for airworthiness when they are operated under stress. If older airplanes like the Stearmans and Waco UPF-7s have been restored and recovered, they are still excellent acrobatic training aircraft. There is nothing scary about performing spins and one (or preferably two) periods of instruction would eliminate the fear and replace it with respect. In addition, a few loops and rolls plus other mild acrobatic maneuvers would improve a pilot's orientation and feel of an airplane. In the event a pilot is later flipped on his back while on final approach by wake turbulence, he would know how to recover without diving into the ground. Mild acrobatic maneuvers performed at a safe altitude in an airplane certified for acrobatics is great fun, and it will instill confidence in the pilot's ability to handle the aircraft and perform maneuvers in all flight attitude regimes.

If we want to train someone to be a bullfighter, we first let them *see* a bullfight. We don't just hand them a red cape and say, "Go through that door out into the arena and see what happens." If you are not skilled in spin recoveries, don't think you're suddenly going to acquire the skill after the airplane enters a spin.

Formal training is the only safe way to get spin training or engage in acrobatic flying, and selection of the flight instructor or flight school is extremely important. If you find an instructor who is not enthusiastic about teaching spins, it would be wise to look for another one with more spin experience. On the other hand, if you come upon an instructor who brags about doing 20-turn spins in airplanes that were only flight tested for six turns, you would be wise in looking still further for an instructor. Attempting to perform spins or other acrobatic maneuvers without receiving proper dual instruction is the least reliable and most dangerous learning method for this type of flying.

FIRST LOOP ATTEMPT

When I was learning to fly, most students were on their own after solo, and good instructors were very scarce. I am relating this story to show the importance of obtaining good dual instruction before attempting solo acrobatics.

The Waco RNF powered by a 125-hp Warner Engine was not dangerously fast, but it was one of the best airplanes that I had the opportunity to fly prior to getting a Private Pilot license. It was a three-place open cockpit biplane, having two seats for passengers in the front with the pilot flying from the rear cockpit (Fig. 10-4). Keeping it polished and selling tickets for airplane rides was the way I earned flying time in the Waco.

One day I heard several pilots talking about doing loops in the Waco, and they explained step-by-step how to do a loop. One pilot warned that if too much back stick pressure was applied in the pull-up, the airplane might snap roll in the top of a loop and this happened to him in his first loop. Many times I had observed loops from the ground but had never ridden in an airplane while one was being performed.

On my next flight in the Waco at 4000 feet above the ground, I cleared the area and prepared for my first loop. Earlier that day, the Waco had been flown on a charter flight and there were some newspapers still on the floorboards that had been placed there to keep oil from getting on a lady passenger's clothing during the trip.

As I recall, this airplane was not equipped with an airspeed indicator. Anyway, the Waco was put into a shallow dive and when the wings and wires started humming "Nearer My God To Thee," the airplane "sounded" like it was ready. The pull-up was started—then I remembered the pilot's warning that it might snap roll by using too much back pressure. Consequently, the

Fig. 10-4. Waco F powered with 125-hp Warner engine.

back pressure was used too sparingly, and I soon realized the airplane was slowing down and getting rather quiet. More back stick pressure was then applied, which caused the Waco to slowly fall over on its back. The engine sputtered and quit, but the propeller kept windmilling. Suddenly, the newspapers blew out of the front cockpit, and I said to myself, "Oh no, the wings must be coming off!" While the Waco was falling upside down, another thought crossed my mind: "I don't believe my safety belt is fastened!" Holding myself in the cockpit by hanging on the stick and throttle was the way this emergency of the moment was handled. As additional back pressure was applied, the nose dropped with a gradual buildup of airspeed as the airplane became alive. After completing the triangular-shaped loop and returning to level flight, the safety belt was checked. Yes, it was still fastened.

Of course, when a maneuver goes sour for an experienced pilot and an incident of this nature is encountered, it is considered a no-big-deal occurrence. However, when a neophyte pilot attempts to experiment with the unknown, it is quite possible that the pilot could run out of knowledge or goof up while under stress. Hopefully, relating this episode of impersonating an aerobatic pilot will help convince inexperienced pilots that they must obtain dual instruction before attempting acrobatic maneuvers. So don't feel that it detracts from your masculinity if you decide not to try solo spins and aerobatics without proper dual instruction.

OBTAINING SPIN AND AEROBATIC TRAINING

Several organized clubs for aerobatic pilots, including the International Aerobatic Club, can provide the names of instructors with outstanding reputations in this field. There are a number of flight instructors and flight schools around the country that specialize in spin training and aerobatics.

In Oklahoma City, there is an aerobatic pilot by the name of Dan Stroud who conducts the type operation of which I am speaking. Dan works in the mornings as news director for radio station KOCY and KXXY-FM. In the afternoons, he teaches spins and aerobatics in his Bellanca Super Decathlon (Fig. 10-5). Eighty percent of his students are commercial pilots, flight instructors, and airline transport pilots (ATPs). The remaining 20 percent are students and private pilots.

Airshow pilot Bob Bishop taught Dan aerobatics and assisted him in setting up a four hour primary aerobatics course and a ten hour course syllabus. The four hour course consists of chandelles, lazy eights, loops, rolls, and Immelmanns with the last two hours devoted to spin training. Spins are entered from over-the-top and out of the bottom of turns, crossed control conditions, etc. Dan starts his spin training at 7000-8000 feet and never permits students to practice them below 5000 feet. Since the Decathlon is equipped with a 180-hp Lycoming engine and a constant-speed prop, it climbs to 8000 feet in five or six minutes. This airplane has a quick-release door-opening mechanism and parachutes are worn at all times. With an inverted fuel system and the symmetrical airfoil wing, the Decathlon flys about as well inverted as it does right-side up.

If you are contemplating getting spin or aerobatic training, I suggest you find someone like Dan, who specializes in acrobatic flight. For further information contact:

International Aerobatic Club, Inc. (IAC)
P.O. Box 229
Hales Corner, Wisconsin 53130
(Division of Experimental Aircraft Association)

Fig. 10-5. Dan Stroud beside his Super Decathlon in which he teaches spins and aerobatics. (courtesy Dan Stroud)

143

Fig. 10-6. Frank Hand (left) and the late Bevo Howard in front of a J-5 Waco Taperwing.

CRAZY FLYING ACT

When I contacted Frank E. Hand, Jr., of Fort Worth, Texas, requesting stall/spin stories, I didn't realize that he had been engaged in "Crazy Flying" airshow acts from 1937 through 1941. At that time, Frank was a flight instructor and Chief Pilot for the late Bevo Howard at his flight operation in Columbia, South Carolina (Fig. 10-6). He often performed dual comedy acts with Bevo, using two airplanes in front of the line of airshow spectators. Frank soloed in the late 1920s in a Waco 9, and I heard that he and Pontius Pilate had the same flight instructor. During World War II he was in the Navy (NATS) flying DC-3s and DC-4s. His time in service with the FAA was 30 years, and when he retired in 1972, he was Chief of the Air Carrier District Office in Fort Worth, Texas.

The following quotes are highlights of what Frank Hand told me about his experiences with stalls and spins:

"Down in the Carolinas, North and South, I did put on a lot of Crazy Flying acts, mostly in Cubs. The Curtiss-Wright Junior, and Taylor E-2, J2, and Piper J-3 Cubs proved to be the most maneuverable aircraft for successfully performing this type of flying. I played the part of the farmer during these crazy flying acts.

"Introduction of the act was usually handled by an airshow announcer conveying this message to the public: A local farmer has just presented us with a book he has written entitled, *How To Teach Yourself How To Fly*. This farmer, Mr. Schildknechty, is seeking permission to give everyone in attendance a demonstration with the use of an aircraft. He wants to show how well his book is written and how thoroughly it covers all aspects of flying.

"At reduced power, the aircraft starts its ground roll, executing 90, 180, and 360 degree turns with the tail up as well as down, dragging wingtips and throwing dust from the tips as it is maneuvered. After a few minutes of this, the aircraft staggers into the air with about half-power from the engine, and proceeds to make 90, 180, and 360 degree turns, some turns banked but most of them skidding at an altitude of 50 to 200 feet AGL (above ground level). Three approaches to a landing are made with bounces and pull-ups. Finally, the airplane manages to arrive at an altitude of 800 to 1200 feet AGL, where several loops and snap rolls are made. The final maneuver consists of stalling the airplane at 800 to 1200 feet, entering a spin, recovering after one and one-half turns and thence a normal landing, followed by a round-house ground loop.

"Goodness knows how many of these I flew, but it was a bunch. I only experienced one close call and that was in an E-2 Cub. I kicked it off in a spin at 900 feet AGL with the intention of doing my usual turn and a half and landing out of it. But this day, after beginning recovery procedures at a turn and a quarter, the E-2 was still rotating with a solid feel of continual rotation. The runway I was spinning down to was really beginning to look mighty wide. I can still remember it. I don't recall doing anything other than applying additional pressure to the controls. Needless to say, when the rotation stopped I was close. Thereafter, I started spins from a higher altitude when performing in the E-2.

"The only flat spinning airplane that I ever had anything to do with was a Fairchild 22 with a Warner 125-hp powerplant. It was a two-place open cockpit monoplane, a beautiful flying little airplane, with really no bad habits. Its spin characteristics were normal with one exception. After a turn and a half the nose would rise to a point just below the horizon and stay there all the way to the ground if you allowed it to do so. The stick would remain all the way back during the spin, requiring about five pounds of forward pressure to move it from that position. The pilot could take his feet off the rudder pedals, fold his arms and watch the world slowly go around.

"Recovery from such a spin attitude in the Fairchild 22 was never any problem. If NACA recovery procedures were used in a normal manner, the aircraft always popped right out and was flying again pronto. In the beginning the CAA/War Training Service was very reluctant to give their stamp of approval for using the Fairchild 22 in the Secondary CPT program due to its flat spin characteristic. They finally relented if we would perform all of the acrobatic maneuvers in the curriculum before a board of witnesses, with spins entered from steep turns, over the top of turns, snap rolls, etc. All spin recoveries had to be made with the pilot's arms and hands extended outside the cockpit to prove the aircraft was recovering as a result of its inherent stability.

"I was elected to fly the flight test and successfully completed the sequence of maneuvers. All spin recoveries were accomplished with my arms fully extended as specified. At Charleston, S.C., we flew this airplane through several secondary CPT programs with great success.

"Military Stearman trainers, T-6s, Waco UPF-7s, and Waco F-2s are the only aircraft I ever spun inverted. In every case these spins were intentional, and circumstances were normal.

"The military still believes in spin training and continues to teach spins throughout their flight training curriculum—and rightfully so. It is a shame that money eliminated the necessity for everyone in civil aviation to learn to fly and be knowledgeable in regard to spins. Charge it up to politics— knocking out the spin requirements and substituting 'awareness' regarding stall-spin happenings, preventive measures, and recovery techniques. This promoted the involvement of many people in flying who later eliminated themselves because they didn't know anything about spins. Today there would be a lot of folks still walking around enjoying life if they had been familiar with this subject prior to the moment the incident/accident occurred. We really didn't promote aviation, did we?"

SUMMING UP

We have gone to a lot of trouble to keep from teaching people to fly—horns blowing, lights flashing, bells ringing, and stick shakers. Some of these are good safety devices, yet there is no substitute for learning the full range of the controls.

Stall/spin accidents usually occur from normal appearing attitudes, such as climbing turns, gliding turns, and steep turns—not necessarily from steep pitch attitudes. Pilots must realize that all airplanes don't have the same gentle stall behavior as the one in which they learned to fly. Stall recoveries overlearned in a small docile trainer and seldom practiced since mean that the pilot will most likely regress to an earlier behavior pattern when an inadvertent stall occurs. What you do in some situations has to be instinctive because you don't have time to think about it, and no one gets ready for an emergency in a moment. I remember reading a statement once made by the United Kingdom National Safety Council that goes something like this: "What a pilot does in an emergency is determined by what he has been doing regularly for a long time."

Pilots have been told for more than 30 years that using ailerons in stalls is not hazardous to your health. I hope I have convinced you this may not be true.

The main points I have tried to stress are:

☐ *If a stall does not occur, a spin cannot occur*.

☐ If you use recovery action soon enough in a stall, you won't need full rudder and down elevator later to recover from a spin.

☐ Place more emphasis on the use of instant, brisk, and positive recovery elevator to recover from *all* stalls.

☐ Refrain from using the ailerons in stall recoveries until the wings are unstalled. Never, repeat *never* use the ailerons in a stall with the yoke in your lap.

☐ Spin training should be a must for pilots flying spinnable airplanes.

The opinions expressed in this book are my own and do not, of course, in any way reflect official views. On the other hand, I have considered these beliefs true and correct for many years, and a large number of people in aviation also share them.

Bibliography

Acrobatics—Precision Flying With A Purpose, FAA AC91-48, USGPO, Washington, D.C., June 1977.

Air Training Command, *Primary Flying,* ATC TP-1111-1, 111128, July 1959.

Air Training Command Manual 51-3, *Aerodynamics For Pilots,* July 1970.

Beechcraft, *Safety Information,* March 1981.

———— *Skipper 77 Pilot's Operation Handbook,* September 1979.

Bowman, James S. Jr., *Airplane Spinning,* NASA SP-83, March 1966.

———— *Summary of Spin Technology As Related To Light General Aviation Airplanes,* NASA TN D-6575, December 1971.

Burk, Sanger M. Jr.; Bowman, James S. Jr.; White, William L., *Spin-Tunnel Investigation Of The Spinning Characteristics Of Typical Single-Engine General Aviation Airplane Designs, I-Low-Wing Model A: Effects of Tail Configurations,* NASA Technical Paper 1009, September 1977.

Cessna, *Spin Characteristics of Cessna Models 150, A150, 152, A152, 172, R172 & 177,* May 1980.

Civil Pilot Training Manual, Civil Aeronautics Administration, 1941.

Cole, Duane, *Conquest Of Lines And Symmetry,* Ken Cook Company, Milwaukee, Wisconsin, 1970.

———— *Happy Flying, Safely,* Ken Cook Company, Milwaukee, Wisconsin, 1977.

Conway, Carle, *Joy Of Soaring,* The Soaring Society of America, Inc., 1969.

Flight Instructors' Handbook—CAA TM 105, Civil Aeronautics Administration, January 1956.

Flight Training Handbook, AC61-21, Federal Aviation Agency, 1965.

Flight Training Handbook, AC61-21A, Federal Aviation Administration, 1980.

Hazards Associated With Spins In Airplanes Prohibited From Intentional Spinning, FAA AC61-67, USGPO, Washington, D.C., February 1974.

Hurt, H. H., *Aerodynamics For Naval Aviators,* NAVAIROO—80T-80, U.S. Navy, 1960.

Lowery, John, *Anatomy Of A Spin,* Airguide Publications, Inc., February 1981.

Piper, *PA 38-112, Tomahawk Owner's Handbook,* April 1981.

Schweizer, *Soaring School Manual,* August 1979.

Syllabus And Study Material From The FAA General Aviation Pilot Stall Awareness Training Study, FAA, 1979.

Special Study General Aviation Stall/Spin Accidents—1967-1969, National Transportation Safety Board, September 1972.

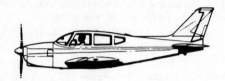

Index

Edited by Steven Mesner